9/27/10

RUNNING OUT
OF WATER

RUNNING OUT OF WATER

THE LOOMING CRISIS AND SOLUTIONS TO CONSERVE OUR MOST PRECIOUS RESOURCE

PETER ROGERS

AND

SUSAN LEAL

FOREWORD BY
CONGRESSMAN EDWARD J. MARKEY

palgrave
macmillan

To Suzanne Ogden and Susan Hirsch

RUNNING OUT OF WATER
Copyright © Peter Rogers and Susan Leal, 2010.

First published in 2010 by PALGRAVE MACMILLAN® in the United States–a division of St. Martin's Press LLC, 175 Fifth Avenue, New York, NY 10010.

Where this book is distributed in the UK, Europe and the rest of the world, this is by Palgrave Macmillan, a division of Macmillan Publishers Limited, registered in England, company number 785998, of Houndmills, Basingstoke, Hampshire RG21 6XS.

Palgrave Macmillan is the global academic imprint of the above companies and has companies and representatives throughout the world.

Palgrave® and Macmillan® are registered trademarks in the United States, the United Kingdom, Europe and other countries.

ISBN 978-0-230-61564-9

Library of Congress Cataloging-in-Publication Data
Rogers, Peter.
   Running out of water : the looming crisis and solutions to conserve our most precious resource / Peter Rogers and Susan Leal.
      p.   cm.
   Includes index.
   ISBN 978–0–230–61564–9 (hardback)
   1. Water conservation. 2. Water-supply. I. Leal, Susan. II. Title.
TD388.R64   2010
363.6'1—dc22

                                                                                2010007929

A catalogue record of the book is available from the British Library.

Design by Letra Libre, Inc.

First edition: August 2010

10 9 8 7 6 5 4 3 2 1

Printed in the United States of America.

CONTENTS

FOREWORD

In the United States, we often take water for granted. A flood or drought might focus our attention, but after the rivers recede or rain falls, we think little about water. For most of us, our most pressing water concerns are not unlike the big one here in Boston: Will Saturday's Red Sox game be rained out?

Yet water is our lifeblood. It touches almost every part of our lives. It affects our health, our economy, and our environment. It's essential to the food we eat and the goods we produce. Every time we buy something—food or a car or a pair of shoes—we are also buying water.

Unfortunately, we have a habit of using water as if there is an unlimited supply. We waste it even though the supply of fresh water is both finite and threatened. More and more we read about water shortages throughout the globe, and increasingly those shortages are close to home. Compound overconsumption with aging infrastructure, pollution, and climate change, and we are facing an unprecedented stress on our freshwater supply.

Like oil, water is a strategic resource, and we must use it efficiently. Unlike oil, there are no substitutes for water. So the need to use water as effectively as possible is essential, which means we cannot afford to contaminate it. And, if we do contaminate our water supply, we must make sure that we clean it up. W. C. Fields is quoted having said, "I never drink water because of the disgusting things fish do in it." These days, we need to be concerned about the disgusting things we are putting into our water and what we are doing to the fish—and, potentially, to ourselves. As Chairman of the U.S. House of Representatives Energy and Commerce Committee's Subcommittee on Energy and Environment, I am well aware of the toxins ending up in our lakes, rivers, and oceans, making our water unsafe to drink or even swim in. During 2009 and early 2010, in hearings before my subcommittee, we heard sobering testimony on the presence of toxic chemicals that are getting in waterways and the potential harmful effects that these chemicals can have on public health and the environment.

IF OUR BATTLE to protect this precious resource were a game—and, sadly, it is not—our team would be trailing with the clock running out. The challenges to protecting this irreplaceable resource are growing: An increasing population will need more water. Then, there is climate change. I have spent many years working with world leaders in government and science in an effort to understand the implications of climate change and explore how we can slow down its effects. That is why I am proud to be a coauthor of the Waxman-Markey comprehensive energy and climate legislation. But, even if we are successful in slowing down global warming, it will still put further strain on our water supplies. The threats of drought, extreme weather events, and accelerated glacier melt are directly impacting our water supply and, ultimately, our environmental and national security.

Water is finite, but fortunately there is no limit to human ingenuity and creativity. This brings me to my good friends from Cambridge, Massachusetts (just down the road from my Congressional District), Peter Rogers and Susan Leal, the authors of this wonderful book you now hold in your hands. In *Running Out of Water* they will introduce you to leaders around the world who are finding creative solutions for conserving and protecting our water supply. These solutions are multifaceted and are taking place on all levels. From tireless and essential work being done by UNICEF and the World Health Organization to progressive and groundbreaking action by individuals in cities and towns, measures are being taken toward mitigating adverse impacts on our water security.

Peter, a Harvard professor of Environmental Engineering and Susan, a Harvard fellow and former head of a large water utility, describe examples of more efficient, renewable water use. In some cases, they describe using new and more advanced methods for removing contaminants from water. In other examples, they share successful stories of turning wastewater into fuel and affordable methods of bringing safe drinking water to slum dwellers in developing countries. They describe leaders in government and business taking ground-breaking action to protect our water. In many cases, a project's success requires the participation and support of everyday people.

The technology is often available to make improvements, and limited actions have begun globally, but the motivation to act on the scale we need is missing. This book can help provide that motivation. It is a call to action as well as a celebration of the progress already underway.

People genuinely concerned with saving the world could well start with saving its water supply. The need is great, the urgency is here, and the time is upon us to save our supply. *Running Out of Water* offers hope and guidance for getting that crucial job done.

—Edward J. Markey
U.S. House of Representatives
Chairman, Subcommittee on Energy and Environment

CHAPTER ONE

TURN ON THE TAP AND OUT COMES THE WATER

Water or oil: which is more essential in powering our society? From brushing our teeth to greening our lawns, from raising cattle to generating energy, our society is critically dependent upon water. In the United States, we're lucky: we turn on the tap and out it comes, clean, quick, and clear, and it's relatively inexpensive. But, whether you know it or not, there is an impending water crisis that will affect every aspect of your life and the lives of your descendants.

You may have heard about some parts of the United States, or far-flung corners of the world, that have recently endured a drought that was more serious than the occasional dry spell. You may have read that a river—perhaps even the source of your own drinking water—was heavily contaminated by runoff from an upstream farm or factory. Yet we tend to take our access to clean water for granted. When we turn on the tap, when we flush the toilet or take a shower, or when we buy groceries, eat a hamburger, or drink a glass of milk, it never occurs to us that we are in fact using an immensely precious resource, water, of which there is a finite quantity on our planet. Running out of water will spell disaster for everyone and everything.

In fact, there is an ever-widening gap between water demand and supply. Between 1900 and 2000, the world's population grew three-fold, but our water use has increased sixfold. And, while oil consumption has also outpaced population growth, there is a big difference between water and oil. We have begun to find substitutes for oil; but there is no substitute for water.

This book is not meant to overwhelm you with scenarios of doom. What's needed is not fear but understanding and community action. Our goal is to offer some hope by supplying possible solutions. Our remedies will highlight some extraordinary efforts and even radical solutions being initiated by forward-thinking governments, businesses, and individuals in the United States and around the world. We will give you a sense of the solutions that are readily available and feasible to take on this problem.

OUR WATER "FOOTPRINT"

It is not surprising that few people understand there is an impending water crisis or know what to do about it. After all, it's unlikely that the tap will be dry when you turn it on tomorrow. But the impending crisis is real and will affect more than the green quality of our lawns. As we discuss in the following case studies, the problem involves both the quantity and the quality of the water on Earth. If we don't change how we use, reuse, and dispose of water, our way of life will be profoundly affected, perhaps most noticeably in the cost or availability of food.

How badly do we need water? How much do we really use? The answers may surprise you. Yes, we need water for such everyday household functions as drinking, bathing, and sanitation. In the United States, we each use an average of about 100 gallons per day for basic household functions. These 100 gallons, however, represent only a small portion of our actual water "footprint"; most of the water we use is spent in producing the food we eat, whether vegetables, grain, or meat. In the United States, our total water footprint is 1,800 gallons per person per day; this is 1.5 times the figure for the developed world, and twice the world average.

> *It takes 53 gallons of water to produce one glass of milk. It takes 634 gallons of water to produce an 8-ounce steak. When you add up our total water footprint in the United States, it is close to 1,800 gallons per person per day.*

Developed nations in particular use an enormous amount of water. Before we look at how we use such large quantities, let's take a look at where water comes from and how much is available.

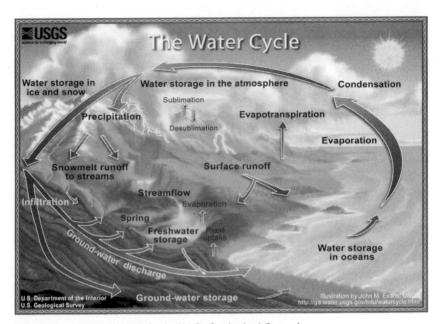

Figure 1.1   Global Water Cycle (*U.S. Geological Survey*)

As this diagram shows, the volume of water on our planet stays constant. It observes a water cycle of evaporation, condensation, and precipitation, changing form but never total volume. According to Earth's water cycle, less than 1 percent of total water is accessible as fresh water. Close to 97 percent of the water is in the oceans and 3 percent is fresh water, but most of the fresh water is inaccessible because it is locked up in ice and glaciers.

As the water cycle illustrates, water is finite. The good news is that water is renewable—it can be reused over and over—but the bad news is that most of us normally use and dispose of water in a nonrenewable and unsustainable fashion. And if we don't change the way we are using water, it will become the "oil" of the twenty-first century. The predicament concerning water is even more critical than that of petroleum. Without water, life cannot be sustained.

Although more than 70 percent of the earth's surface is covered by water, only 3 percent of all water is fresh water, and less than 1 percent is actually accessible to us. The total volume of water on earth never changes—only the form (rain, snow, ice, and so on).

A FINITE WATER SUPPLY STRESSED BY POPULATION GROWTH AND AFFLUENCE

We have a tiny amount of fresh water—less than 1 percent of all that's available on the planet—to share with the other 6.7 billion inhabitants of our planet. While the volume of water never changes—it hasn't changed since prehistoric times—the number of people who need it is constantly growing. The world's population is expected to grow to 7 billion by 2012 and to more than 8 billion by 2025. By then the population of the United States alone is expected to increase from the current 304 million to 357 million.

Along with the planet's growing population, there has been a simultaneous, widespread rise in the standard of living, which increases water consumption. Affluence drives up consumption, and the United States has been leading the way with the largest water footprint. As the better-educated and higher-paid populations of China and India ascend into the middle class, they are increasing their water footprint, too. China, for example, has increased its consumption of beef—a water-intensive food—by 38 percent from 2001 to 2006.

Even at our current population levels, we are facing shortages in our water supply. Across the globe, water problems range in severity: some experience intermittent shortages—36 states in the United States have some water supply problems; in some parts of the world, there is little or no access to safe drinking water; and in many countries, women and girls spend several hours a day traveling for miles to gather water for their families.

Water use increased with population growth and was further driven by affluence: from 1900 to 2000, as world population grew threefold, water use increased sixfold. By 2000, agriculture consumed 70 percent of the total water used globally, with cities and industry consuming the bulk of the remainder.

You may be thinking there's a simple solution: we can just desalinate ocean water. But it's not quite that easy. Desalination uses a lot of fossil fuel energy and is very expensive. Farmers in California normally pay less than $50 per acre-foot for water; desalinated seawater may cost $700 or more per acre-foot. At those prices, we will pay a lot more for our food. Aside from the production costs, there have been environmental problems—in particular, problems with disposal of brine—experienced with desalination. So, this seemingly obvious remedy offers only a partial solution to the quantity of water available.

IT'S QUALITY AS WELL AS QUANTITY

Our finite amount of water is also stressed by increasing household and industrial contamination. Every time we use water, we also have to dispose of it as wastewater; most often, we deposit that wastewater into an ocean or return it to a lake or river. And the more we use, the more we dispose of. An important consideration is ensuring that the water we use is properly treated and cleaned before being disposed of. In some cases, which we will discuss in depth later, we have gone beyond a simple act of treating and disposing of water—by filtering it, cleaning it to a higher quality, we can use it again and again.

In developing countries, overstressed or nonexistent wastewater disposal systems create public and environmental health hazards. Around the globe, there have been efforts to improve supplies of safe drinking water, but they often must contend with contamination from sewage. The second leading cause of preventable deaths in children

Recent cases show that even in affluent, developed countries, people are susceptible to poisoning from human or industrial waste unless they are vigilant in monitoring their wastewater disposal and water treatment.

throughout the world is a lack of access to safe drinking water, or exposure to contaminated water, most often rendered impure by the scarcity of proper human waste disposal systems. In India, for example, in the slums in several big cities, there is a constant struggle to eliminate the contam-ination of water that is caused by human waste dumped in the streets.

Developed countries are not immune, either. In April 1993, failure to properly treat water that had been contaminated by animal wastes from Wisconsin's famed dairy farms resulted in 400,000 people in Mil-waukee becoming ill (half the population served by its water system, with 4,000 hospitalized and 54 deaths). These cases show us that un-less we are vigilant in monitoring how we dispose of our wastewater and treat our drinking water, we can easily poison ourselves and the surrounding animal habitat.

CLIMATE CHANGE AND THE CONNECTION BETWEEN WATER AND ENERGY

Climate change adds another layer of complication by weakening our freshwater supplies and impacting the operations of our wastewater systems. The gradual rise in global temperature is threatening to re-duce the water supply in California, one of the largest food-producing regions of the United States, by 25 percent. On the East Coast, the nation's largest water utility, the New York City Department of Environmental Protection (DEP) is feeling the effects of climate change. Past climate variability and extreme weather events presented challenges to the DEP's water supply and wastewater systems and have guided their initial understanding of the need for climate change adaptation strategies. "The timing and extent of climate change is uncertain and modifying large-scale infrastructure systems is expensive and takes time, but the DEP is committed to minimizing the risks and understanding the challenges of these effects on our water systems," said Angela Licata, Deputy Commissioner, New York City Department of Environmental Protection. Around the world, recent weather patterns appear to be increasingly unpredictable, with more severe storms, spring flooding, and longer periods of drought. Even if we can't be sure of the time frame or the severity of the effects of climate change, we can be sure that it will be a real wild card in planning for our future water sources.

The relationship between water and climate change presents a catch–22. The water world is caught in a vicious cycle: Climatic change is reducing our water supply, and the systems we use to deliver clean water and treat wastewater produce the same greenhouse gases that contribute to climatic change. To understand the connection between climate change and water, it is important first to understand that there is a connection between energy and water. In the United States and most of the developed world, water treatment, the process of cleaning the water that comes from lakes and rivers to make it potable (drinkable), takes about 1 to 2 percent of the nation's total energy supply. Wastewater treatment, the process of treating sewage (cleaning water after we use it) so that it can be safely discharged into rivers or lakes, consumes another 1 to 2 percent of our energy use. The U.S. Geological Survey estimates that this roughly 3 percent of energy produces 45 million tons of carbon emissions every year. Just supplying water in some regions uses significant amounts of energy. California, for example, uses large quantities of energy to move water from its sources in Northern California to consumers in the southern part of the state. Expenditures of resources and their environmental consequences may not have seemed so large in the early days of lower populations with little concern about greenhouse gases and less expensive energy, but now the costs are considerable and growing.

Moving water from Northern to Southern California is the state's single greatest use of energy. California uses 19 percent of its overall electricity, 32 percent of its natural gas, and 88 million gallons of diesel per year in water-related energy use.

The relationship between climate change and water extends into hydropower, which generates up to 6 percent of U.S. energy and 20 percent of energy worldwide. Hydropower has long been considered a clean source of power, and is important in reducing greenhouse gases. But, as water supplies diminish or change seasonal attributes due to climate change, less hydropower will be produced. This reduction in energy will, in turn, likely be replaced by other methods of producing energy, such as coal-burning plants, that yield even more greenhouse gases. It's a vicious cycle that can be mended, as we will show in the following chapters.

WHO IS IN CHARGE OF YOUR WATER?

We don't often think about where water comes from, but for water to reach your tap, someone is responsible for getting it there. As we confront interrelated issues of water supply, water pollution, and climate change, who have we designated to deal with these issues?

Someone has to make sure that there is an adequate water source with a well-maintained system of reservoirs, tunnels, and treatment plants—ensuring it is safe to drink—delivering water to people, industries, and agriculture. Someone has to make sure that after water is used, the resulting wastewater is properly treated, not just dumped "raw and dirty" into the nearest lake, river, or ocean (see Figure 1.2

on pages 16–17). For all these efforts and more, there are government-run or -sanctioned water utilities or agencies charged with running well-maintained systems and supplying water for the next twenty to thirty years (and, we hope, for many years beyond that). How well these agencies do their job will have a profound effect on our quality of life.

The people in charge of our water supply and wastewater treatment are often elected or appointed officials who are not visible to the average citizen, and how they get into these positions of power varies by city, state, and country. In New York City, the head of the Department of Environmental Protection is appointed by the mayor. In Orange County, California, the nine-member water board is elected by the voters of Orange County. And, to complicate matters, state legislatures can often pass legislation that may determine how water is to be used. At the federal level, too, there are over twelve federal agencies responsible for one aspect or another of water delivery or sewage treatment. Around the globe, there are a variety of government and private companies, often operating as a concessionaire of the government, that may have a say in how water is provided. In some cases, those in charge have experience in managing water or running government systems, but, unfortunately, there are plenty of instances where the appointment is made solely because of political or personal connections.

One of our aims in this book is to identify some of these decision makers and explain how they can make a difference in protecting this precious resource for generations to come. We will provide suggestions about how you as a consumer can influence them to protect this precious resource.

HOW CAN WE CONFRONT
THIS LOOMING CRISIS?

The purpose of this book is to explain the predicament we face and to shed some light on possible remedies. We will not spend time explaining the benefits of taking shorter showers or installing low-flow toilets, although those are good ideas. Instead, we will give you the big-picture solutions, ideas that are currently making an important difference around the globe. Our hope is that people will begin to recognize these solutions as affordable, sustainable practices they can replicate in their own communities.

We will explain the ins and outs because we have been involved with water from both the inside and the outside. One of the authors, Peter Rogers, is a Professor of Environmental Engineering at Harvard University. He has studied water access on many continents, served as adviser to governments and international organizations, and trained water management engineers and economists. The other author, Susan Leal, has seen water management from the inside as head of a large utility providing water to close to 2.5 million people in Northern California. It is our hope that, together, our perspectives will give you a more complete picture of both the water problems facing the world and an array of workable solutions for fixing them.

As a professor and a practitioner, we have expertise in the technology available to take preemptive steps against a water shortage. As we describe some of the solutions, we will explore how politics, science, and economics affect delivery of water in several locations around the globe.

We believe the crisis can be solved—we couldn't have written this book if we didn't. To demonstrate how the predicament can be addressed, we will show you how numerous communities, and, in many cases, determined individuals, have taken political risks and fought social conditions to deal with the challenges of water deficits and other regional or local water crises.

In Singapore, for example, leaders worked hard to gain public acceptance before embarking on a project to transform sewer water into drinking water. San Francisco's water utility reached out and engaged the city's residents and earned their financial support to retrofit its aging sewer system. In yet another success story, Brazil's leaders and private contractors were able to provide necessary sewers to the urban slums through the use of the "sweat equity" of the local residents.

> *Technology is available to solve our current water crisis. The key ingredient is action.*

You might wonder how a problem with scientific or technical solutions could be up for debate or allowed to drift, but public policy often depends on how citizens and elected officials think about such things (or don't think about them, as the case may be). For example, the federal government estimates that every year the national investment in sewer infrastructure falls short by more than $20 billion. And yet, even though the public health and the environment are in jeopardy, we have not made the necessary investments.

Just as it took a massive campaign to get people to stop smoking or to move away from buying gas-guzzling cars (both only partial successes), we must realize that we need to change how we manage our water resources. Change will of course depend on public cooperation. Actions often hinge on ratepayers' willingness to pay more or on public approval of government bond measures to pay for improvements. Generally speaking, change will depend on winning public acceptance of more efficient ways of using water, just as the people of Singapore, San Francisco, and Brazil have done.

It is our hope that you will come away from this book with a desire to be more informed and engaged in future water security, both locally and globally. And we hope you will be motivated to demand more action (and less inertia) from your leaders. We hope you will see that there is indeed a water crisis on the horizon, and that there's genuine hope for solving it.

How Water Gets to Your Tap and Leaves Your Home or
Business—the Process for Delivery of Drinking Water
and Disposal of Wastewater

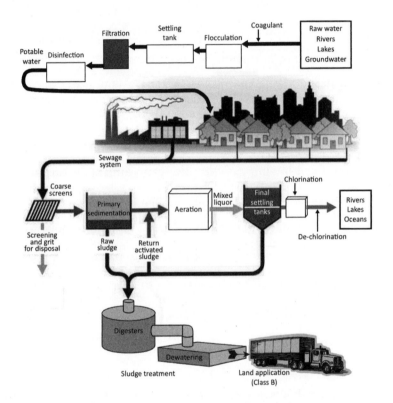

Figure 1.2   In most developed countries, water and wastewater go through a
treatment process that is similar to the one diagramed here.

Water Treatment Process (upper tier of Figure 1.2)
Raw water comes from freshwater sources, such as lakes, rivers, or groundwater, and is put through a treatment process before it is delivered to homes and businesses. The treatment process includes: flocculation, which is mixing some chemicals (iron salts, alum, or similar compounds) with the water to cause clumping or aggregation of any particles in the water. This allows those particles to settle out as it flows through the settling tanks. Remaining particles that have not settled out will be removed by the filtration process, which clarifies the water and prepares it for the final step, disinfection. In the final step the water is disinfected, most commonly with chlorine or chloramines, to kill the pathogens before it enters the distribution system (pipeline) and is delivered to us.

This is a basic water treatment system; some treatment systems may include additional steps to assure water safety. Municipal water systems in the United States must test the quality of the drinking water tens of thousands of times annually.

Wastewater Treatment Process (lower tier of Figure 1.2)
Wastewater treatment is more involved than the water treatment process, and must go through several steps after it is collected from businesses and homes and before it is discharged into receiving waters, ocean, lakes, or rivers. The treatment begins with the waste going through a screening process to remove items that cannot get through the treatment process, such as wood, sticks, and plastics. The wastewater then moves through primary sedimentation tanks, where the solids (organic matter) settle. In the tanks, the scum on top of the water, which usually contains grease, oils, or soap, is skimmed off. The solids and the scum are pumped out of the tanks and go through the digestion process. In the digestion process, solids are kept for 20 to 30 days in large, heated enclosed tanks or "digesters." These digesters are like concrete stomachs, where bacteria breaks down or "digests" the solid, reducing its volume and odors, and eliminating organisms that can cause disease. The finished product from the digesters is usually sent to landfills but also can be further processed and used as fertilizer. For the remaining wastewater, the process continues with the wastewater going through an aeration process. As organic matter decays, it uses up oxygen and aeration replaces the oxygen. The waste goes through the final settling tanks where the wastewater is disinfected with chlorine to eliminate bacteria. The chlorine is mostly eliminated as the bacteria is destroyed, but often the chlorine is further neutralized so as not to harm other aquatic life. The remaining wastewater or effluent is then discharged into the receiving waters.

CHAPTER TWO

MAKING IT LAST

Using Technology to Recycle Water

We know that the volume of water on this planet is limited. But, with wise management of our water resources, we can make it stretch further. It is not always easy but, as the following stories indicate, it can be done. In this chapter, we chronicle how three different water utilities—two in the United States and one in Asia—responded to different and continually evolving political, ecological, and demographic situations to implement groundbreaking programs.

In our first success story, we look at how a California county, in an effort to avoid water shortages to its 2.3 million residents, had to battle not only increasing costs and urgent environmental issues but also the state's intricate and sometimes bizarre water politics. This illustrates how the county came out ahead in the battle for water in California—and how the state benefited as well.

WATER POLITICS OF THE GOLDEN STATE: NORTH VS. SOUTH

California's challenges have made the word "water" a hot-button, kitchen-table topic throughout the Golden State. Although politics permeates water discussions around the world, California treats water politics like a blood sport. Divisions are deeply entrenched, water is scarce, and yet, the state's economy depends upon clean and even access. Since water was first used in prospecting for gold during the Gold Rush, water has been tough business. Mark Twain weighed in with an accurate and caustic assessment: "In the West, whiskey is for drinking and water is for fighting."

California's water politics date back to the 1840s, when miners constructed wooden flumes—wooden chutes—to deliver water up, over, and around the mountains. Water is not naturally abundant throughout the state. In some areas, primarily Southern California, rainfall is very limited. The technology to move water grew, eventually being applied in the early 1900s to facilitate agriculture and urbanize first San Francisco and later Los Angeles. Throughout the twentieth century, California built a vast and mighty network of water conveyance infrastructure—pipelines, canals, and tunnels—that moves water to both inland agriculture and the coastal cities' growing populations.

TABLE 2.1   SETTLEMENT OF CALIFORNIA'S LARGEST CITIES, SHOWING
THE MIGRATION TO SOUTHERN CALIFORNIA

Largest Northern California cities: San Jose, San Francisco

Largest Southern California cities: Los Angeles, San Diego

	1900	1910	1930	2000
San Francisco	342,000	506,000	634,000	776,000
San Jose	21,000	29,000	58,000	895,000
Los Angeles	102,000	576,000	1.2 million	3.6 million
San Diego	29,000	74,000	147,000	1.2 million

Unfortunately, the natural location of California's water does not equitably coincide with the state's population or its agriculture centers. While about 75 percent of the water is located in Northern California, an almost equal amount is used by the southern half of the state. Thus, Californians face the constant challenge of moving the water from its source in the north to the tap in the south. This transfer of water is an ongoing battle that guarantees a lifetime of employment for lawyers and lobbyists.

A Growing Population Creates a Growing Demand

Compounding California's water problem is a growing population with a corresponding demand for water. California's 2009 population of 38 million is expected to grow by 15 million people by the year 2020, while water supplies remain static. Yet every new building construction or business expansion creates a need for more.

Orange County is a prime example of a growing population that brings with it an increasing thirst for water to be imported from the north. The county's population grew from 704,000 in 1970 to over 2 million in 2000; it is expected to reach 2.4 million by 2020. In the following example of wise water management, we'll see the challenges facing Orange County and how they are being addressed.

Orange County, best known as the home of Disneyland, the late John Wayne, and the subject of many television series, is also noteworthy for protecting its water security: it has followed a very progressive, almost radical, path not yet implemented by any other region in the nation. By recycling its water, essentially converting waste into a resource, Orange County reduced its dependence on Northern California, improved its bottom line, and became a pioneer in municipal water use.

Environmental Problems Threaten Continued Water Flow

In addition to a growing population, Orange County had mounting environmental concerns. Water that travels from Northern California to Southern California must pass through an area called the Sacramento–San Joaquin River Delta, known throughout the state as simply "the Delta." The Delta, a unique ecosystem, is where the snowmelt from the snow-packed mountains feeds into several Northern California rivers and comes together to form a confluence. Water flowing into the Delta is then pulled by massive pumps that lift thousands of tons of water per day up 244 feet into several canals and then downstate to over 25 million Californians for industrial, agricultural, and residential uses. At the southern edge of the San Joaquin Valley, water must be pumped up again over the Tehachapi Mountains, an elevation gain of 1,925 feet, and then on to the Greater Los Angeles metropolitan area, including Orange County.

The Delta supports an extremely precarious ecosystem. Over the past four decades, beginning with the water projects of the 1960s, water pumping has caused severe environmental damage. Most recently, the controversy involved the large pumps pulling water from the Delta, endangering the Delta smelt— a three-inch fish unique to this region— and reducing its numbers. Scientists and environmentalists claimed that the large pumps are sucking up and killing the smelt, and in 2007 the federal courts agreed with the environmental community, which led to restrictions on the amount of water pumped from the Delta.

The Delta is the heart of California, pumping its lifeblood of water for all levels of production and consumption.

Another challenge to the area's precarious ecosystem is that the farmlands built upon islands in the Delta have accelerated the decomposition of the peat soils, resulting in massive subsidence and endangering the entire network of nearby waterways.

In a 2000 report, the U.S. Geological Survey called the Delta the "sinking heart of the State."

Moving Water North to South: The State's Largest Energy Users

In addition to the consequences described above, there are environmental implications in the energy required to move water throughout the state. In fact, California's largest energy users are the water agencies responsible for conveying water from north to south.

Moving water from Northern to Southern California is the state's single greatest use of energy. California uses 19 percent of its overall electricity, 32 percent of its natural gas, and 88 million gallons of diesel per year in water-related energy use.

Millions of tons of greenhouse gases are produced as a result of the fossil fuels used in water-related services. The use of energy and production of greenhouse gases are particu-

Figure 2.1   Heavy dark lines on the map show major aqueducts and the location
of the Sacramento–San Joaquin Delta. Runoff from the snow-packed mountains
feeds into several Northern California rivers and comes together to form the
waterways of the Delta. Water from the Delta travels through the aqueducts to
serve 25 million people and agricultural regions. (*Courtesy of the State of
California*)

larly alarming in California, where the effects of climate change are already apparent, and the effects on the state's water supply are predicted to be devastating.

High Cost of Importing Water

Monetary cost is another major concern. It is amazingly expensive to move water against gravity. Every day, millions of gallons of water are moved hundreds of miles from the Delta to Orange County, and, as with other counties that rely on imported water, Orange must pay the cost of transporting its water.

Orange County also imports its water from another source: the Colorado River. Moving water from there has many of the same drawbacks as importing water from Northern California. To complicate matters, the water of the Colorado is divided among seven states and Mexico. This formerly reliable source, however, has been drying up due to drought and agricultural diversions: a river that once flowed into the Gulf of California, also known as the Sea of Cortez, is now barely a trickle by the time it reaches the sea.

Dilemma: Meeting Orange County's Future Water Demand

Regardless of logistics and costs, when Orange County's 2.3 million residents turned on the tap, they expected a quick and clean flow of water. The management of the county's water utility had to ensure such a flow. However, as the county's water officials assessed their current usage and looked at projections for the next twenty to thirty years, they became concerned about their heavy reliance on Northern California and the Colorado River. Increasing the amount of water imported would subject them to high economic, political, and environmental costs. And they couldn't guarantee that an increase would satisfy the water demands of Orange County and leave enough water for the rest of the state's people, industries, and ecosystems.

In the mid-1990s, Phil Anthony, one of the elected officials charged with managing Orange County's water procurement, saw the situation this way: "The risk (for meeting the county's water needs) was just too big to depend on more water imported from the Delta or the Colorado." He and the eight other directors of the Orange County Water District began to look for ways to improve their access to local sources of water. Orange County's local source of water is its groundwater basin, a large underground reservoir fed by water from both surface and underground waters. Over the years, as the population grew and demand increased, the groundwater basin was being emptied faster than it was being replenished. A five-year drought had taken its toll and, to make matters worse, saltwater had begun seeping into the groundwater basin from the coastline. (As the waters in the groundwater basin are depleted and fall below the level of the adjacent ocean, the seawater enters the basin through force of gravity.) Too much salt would make the water in the basin undrinkable.

Phil Anthony and the other water utility directors looked for help from the Orange County Sanitation District. The sanitation district is tasked with treating and disposing of wastewater from the 2.3 million residents of the county, which is quite a job: for every ounce of water used, another ounce must be treated and disposed of, usually into a river, bay, or ocean. Phil Anthony determined that this "sewer water" should not be wasted, but should instead be cleaned and reused to supplement their local water supply.

The Orange County method for recycling sewer water is through an elaborate three-step process involving *microfiltration, reverse osmosis,* and *ultraviolet light* treatment. One of the key components is the reverse osmosis (RO). In the RO process, water is forced through a membrane with holes so small that pesticides, pharmaceuticals, and viruses—which have much larger molecules—are filtered out.

After the sewer water goes through the three-step process, it is safe to inject it into the groundwater basin, where it mixes with Orange County's other water sources, including water from the Delta and the Colorado River. Then, the mixture goes through the drinking water treatment process (see pages 16–17.) Surprisingly, the cleanest water going into that drinking water treatment plant is the recycled water. Experts who understand the science behind recycling agree that consumers' fears about the purity of recycled water are not warranted.

Let's take you through the three-step process:

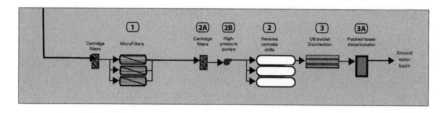

It starts with Microfiltration (Step 1).
Microfiltration (1) is a purification method commonly used to purify baby food and sterilize medicines. The microfiltration process uses straws with tiny holes in the sides that are 300 times smaller than a human hair. By drawing water through these holes, all bacteria and viruses are filtered out.

It continues with Reverse Osmosis (Step 2).
This step begins with the water moving through the **cartridge filters** (2A) which take out any remaining particles before the water goes on to the **reverse osmosis units** (2). Reverse osmosis or RO is one more of the unique technological advancements in the last few decades. The RO process uses **pressure** (2B) to force water with all its impurities through a very fine, plastic sheet-like membrane. The impurities are held back by the membrane allowing only the pure water to pass through to the other side, resulting in purified, near-distilled quality water. It is called reverse osmosis because in the normal osmosis process there is no pressure applied to move solution from one side of a membrane to another.

It finishes with Ultraviolet Light (UV) (Step 3).
If any impurities remain after Step 2, the water is blasted with hydrogen peroxide to eliminate any trace contaminants, and is then treated with **ultraviolet light** (3). The UV process can be compared to exposing water to concentrated sunlight and is similar to the way in which hospitals and dental offices sterilize instruments. In the UV system, the water is exposed to several cylinders containing 144 high-powered lamps. By the time the water has passed through the UV system, it has been exposed to 3800 high-powered lamps. After this three-step process, the water goes through a **decarbonator** (3A) which acts to aerate the water and stabilize its pH level. Then finally, the water is directed back into the County's groundwater basin.

Figure 2.2   Three-Step Recycling Process

Consider the words of the utility manager from neighboring San Diego County, Maureen Stapleton: "Concerns about recycled water are ludicrous. Do people think that the water from the Delta or the Colorado hasn't been used before? They've got to be kidding." This assessment is right on the mark. Whether we realize it or not, for decades treated sewage has been discharged into lakes and rivers that supply our drinking water, including the Great Lakes, the Mississippi and the Colorado rivers, and many of those in Northern California.

There were practical reasons why the Orange County Sanitation District was happy to share its sewer water with the county's water district. With the area's increasing population, the sanitation district's sewage treatment plant and its ocean outfall pipe—a pipe 10 feet in diameter that disperses the sewer water into the ocean—were nearing capacity. Unless some of the sewer water was recycled, the sanitation district would need to enlarge its treatment plant and build a new ocean outfall, expansions that would cost millions of dollars and raise objections from environmentalists. Orange County has more to offer its residents than simply its proximity to Disneyland. Among the premier attractions are the beautiful beaches that are cherished and protected by well-organized advocacy groups. Some of those groups were among the strongest proponents of the recycled water project because they saw the benefits of filtering and recycling water instead of building a new large pipe to dump it into the ocean.

Orange County Delivers

The recycled water program became a winner for Orange County. As the county reduced its dependence on imports from Northern California and the Colorado River, the residents got a more reliable source of water, and the project's up-front costs of $480 million are being offset by savings to the water and the sanitation districts. For the water district, the cost of recycled water is about two-thirds the price of imported water and about one-quarter of the cost of desalinated seawater. (As we pointed out in Chapter 1, desalination uses a lot of fossil fuel energy and is very expensive.) The project saved the sanitation district from the expense of having to expand its sewage treatment plant and build an expensive outfall pipe—not to mention the legal fees that likely would have been incurred to counter environmental opposition to the expansion.

The recycled water program has many environmental benefits, including less wastewater being pumped into the ocean, and the recycled water process uses only half the energy it would take to transport the water from the Delta. Additionally, Delta water not used by Orange County is now available for other communities.

In early 2008, Orange County received approval from state health authorities to begin producing 70 million gallons of water a day from its wastewater. It would be enough water for 500,000 residents of Orange County.

Orange County is way ahead of its Southern California neighbors in reducing its dependence on dwindling and expensive imported water, and is setting a good example for San Diego and Los Angeles. In 2008, several environmental groups sued the city of San Diego for increasing the amount of wastewater being dumped into the ocean. As part of the settlement, the city agreed to reduce the amount of ocean discharge by recycling some of its sewer water.

Los Angeles is now implementing a water recycling program. Due to a worsening drought, climate change, and the environmental restrictions on the pumping of water from the Delta, by 2008 Los Angeles had moved forward with a recycled water project of its own.

Meanwhile, Orange County and its water district have received a variety of awards and recognition from around the world. It must have been an especially rewarding moment for the citizens of Orange County when the city of Los Angeles (however begrudgingly) tipped its hat to its neighboring county. A recent op-ed columnist in the *Los Angeles Times* gave this compliment to Orange County: "It embarrasses me to say it, but L.A. has to learn from Orange County. Behind the Orange Shower Curtain, a massive recycled water system went online in January [2008]. By next year, it'll be reclaiming 70 million gallons a day, cutting the sewage dumped into the ocean and saving millions by not having to buy water from elsewhere."

Orange County's groundbreaking action continues to lead the nation in sound water policy. It served as an example for our next case, the island state of Singapore.

SINGAPORE NEWATER:
WHAT IS THE PRICE OF PEACE OF MIND?

With about a million more residents than Orange County and even less land area, Singapore is an independent country of 5 million people living with a booming economy. But, while its people are relatively affluent, the country has always been water poor. It is twelve hours by plane from Orange County to Singapore. Although physically distant, both metropolises have embraced recycling sewer water as a way to solve distinctly different water supply problems. Unlike Orange Country, Singapore does not suffer the specter of drought. In fact, it is blessed with over three feet of rainfall per year, which should be a very substantial supply for the city. Unfortunately, the tiny island state's lack of space means that there is hardly any available land on which to store its water.

Currently, most of Singapore's water is supplied by its neighbor on the nearby mainland, the Malaysian state of Johor. Like Singapore, Johor has plentiful rains. But, unlike Singapore, it has space for water storage. Johor and Singapore are separated by a short, two-mile causeway that holds large pipes carrying water from Malaysia to Singapore.

Singapore takes a chance on recycling its sewer water to secure its most fundamental resource: water.

This causeway is the only close connection between Malaysia and Singapore, and the relationship between the two countries can best be described as guarded. Because of political differences between them, Singapore is concerned about the security of the water it imports from Johor.

For a brief period in the histories of the two countries, they actually merged together. In 1963, at the end of 144 years of British rule, Singapore became part of the Federation of Malaysia. Within a year of the merger, however, there were two deadly riots in Singapore, in July and September 1964. The riots, fueled by racial and religious tensions (the mainland of Malaysia is primarily Islamic and Singapore is primarily Buddhist), resulted in hundreds of injuries and 40 deaths. In late August 1964, the members of the Malaysian parliament voted to expel Singapore from the Malaysian Federation. Singapore had gained its independence against its own will.

Before this happened, however, an agreement had been put in place to provide Singapore with water from the mainland, specifying that Johor was to supply water until 2061. After Singapore's expulsion from the Malaysian Federation, Johor offered to continue honoring the water contract. But relations between the island country and the mainland remained guarded. For Singapore, there was the possibility that Malaysia could use threats of cutting off the water supply as a way to apply political and economic pressure.

In addition to the political and religious differences, there are economic differences. Over the past three decades, Singapore has become one of the wealthiest nations in the world, with the fifth strongest per capita GDP at approximately $52,000 per person. Malaysia, on the other hand, has a much lower per capita GDP, at approximately $14,000 per person. Singapore is concerned that while the contract expires in 2061, the contract price—at the reasonable rate of one cent per cubic meter—is set to expire in 2011, at which point the water contract could be used to extract significant monies or other concessions from Singapore.

Rather than remain at the mercy of an inse-
cure water supply, Singapore opted for a more
expensive but ultimately more reliable source.
It used a solution that has been chosen by cities
in more arid regions: fully treating, purifying,
and reusing its wastewater.

*For Singapore,
water security is
critical in the face
of political and
economic turmoil.*

The recycling movement, often disparag-
ingly called "toilet to tap," has been greatly aided by the technologies
developed over the past twenty years. Singapore started considering
the possibilities of reusing wastewater in the 1970s, well ahead of most
other water utilities or municipal governments. But around that time
the technologies and high costs did not make recycled water a viable
option. In 1998, Singapore tried again, and by the end of 2002 it had
completed construction of two small recycling plants that produced
less than 1 percent of the water needed for the island state. The treatment method Singapore used, almost identical to that used by Orange
County, includes the three-step process of microfiltration, reverse osmosis, and ultraviolet technologies.

Amazingly, Singapore's water authorities found that the water produced at these first two plants not only exceeded the quality of the
potable water supplied by their water utility, the Public Utilities Board
(PUB), but it also met the drinking water standards of the United
States Environmental Protection Agency and the World Health Organization. Based on this successful demonstration, the PUB has

moved ahead rapidly to expand its capacity to turn wastewater into drinking water. The purified sewer water is called NEWater.

Up until 2009, most of the NEWater was used for industrial applications, but Singapore has begun to add more recycled H_2O to its drinking water portfolio. It was projected that by 2010 NEWater would meet 30 percent of Singapore's water needs.

Singapore is paying much more for the NEWater—the cost is 59 cents per cubic meter versus the cost of one cent per cubic meter under its contract with Johor, but they have attained two key benefits, namely, water security (no longer must they depend on outsiders) and highly purified water that can be used for their high-tech industries.

NEWater is higher quality than Singapore's drinking water, so it is prized by high-tech industries—such as the semi-conductor industry—that require highly purified water.

Singapore's concern over water security is still deeply ingrained in the nation's collective psyche, as expressed by the "founder of the nation," Lee Kuan Yew, former prime minister. At the ceremony for the Lee Kuan Yew Water Prize in 2009, he said:

When Japanese troops invaded Singapore in 1942, one of the first things they did was to blow up the pipes transporting water from Johor to Singapore. This left the British colonial army and Singaporeans with only two reservoirs of water that could last two weeks—at most.

More than sixty years after the horrors it experienced in World War II, the tiny island country finally has water security for its people.

IN THE NEXT CASE STUDY, we will focus on one of the pioneers of water recycling. With the passage of the federal Clean Water Act of 1972, which set new standards for water and wastewater treatment, the federal government gave sizable grants to several municipalities to encourage them to upgrade their sewage treatment to meet the new federal standards. St. Petersburg, Florida, used these federal monies to help them implement the first water recycling plant in the U.S. in 1977 with an expansion in 1990.

EARLY BEGINNINGS:
BLAME THE AIR CONDITIONER

St. Petersburg, Florida, a small resort town on the peninsula between Tampa and the Gulf of Mexico, grew from a population of 300 in the early 1890s to a mid-sized city with a quarter million people. (The St. Petersburg-Tampa-Clearwater metropolitan area's population is about 2.4 million.) St. Petersburg began, in the late 1800s, as a resort town for "snowbirds"—tourists and seasonal visitors who wanted to escape the cold and icy northern winters. The city's population experienced significant growth after 1914, when a major league baseball team established its first spring training camp in St. Petersburg. The population continued to grow with the post–World War I proliferation of the automobile; snowbirds from the North began to flock to St. Petersburg by both car and rail during the winter months. But the real boom came in the 1950s with the advent of household air conditioners: the populations of St. Petersburg and other Florida towns began to balloon as formerly winter vacationers became year-round residents.

Like many cities with an expanding population, St. Petersburg began to experience water shortages. The city's increased demand began to outpace its water supply by the late 1960s. The city took visionary action to respond to its water shortages: St. Petersburg was the first American city to recycle its sewer water and use it for landscape irrigation. These efforts to produce additional sources of water were cutting-edge and allowed the city to meet many of its water demands. The only drawback from its recycled water success was that there was less motivation to move away from water-thirsty landscaping.

In Tampa-St. Petersburg, there is not enough water for winter snowbirds to become year-round residents.

For decades, St. Petersburg had drawn its water from groundwater wells in the Cosme-Odessa area located ten miles north in neighboring

Hillsborough County. By 1970, with a growing population and increased pumping of water from the Cosme-Odessa Well Field, there was a drop in water levels of the lakes in Hillsborough County. And not only were the lake levels dropping, but saltwater was beginning to be detected in some of those underground wells. Saltwater intrusion was caused by the underground wells being so depleted that they could not keep the seawater from the nearby Gulf of Mexico from seeping in. As a result, St. Petersburg was forced by Florida's state water authority to decrease by half its pumping of the Cosme-Odessa wells. In response, St. Petersburg began drilling wells and drawing water from wells farther north of the city and exploring other ways to supplement its water supply.

St. Petersburg spent the next several years looking for solutions, ultimately deciding to recycle its sewer water and apply it to the greatest portion of water use: landscape irrigation. In 1976, St. Petersburg began building the first municipal recycled water plant in the nation. The biggest concern was whether the consumers would accept the recycled water.

St. Petersburg went forward with its plan using the existing technologies—which did not yet include the reverse osmosis process—for treating or "cleaning up" the sewer water for reuse. The result was recycled water of a quality that was far from drinking-water standards. The health authorities ordered that the recycled sewer water could only be applied by water irrigation professionals and was not suitable for use in residential landscape applications. And, because this recycled water could not mix with drinking water, St. Petersburg had to build a network of pipelines separate from its drinking-water distribution system in order to deliver recycled water for industrial and commercial applications, such as golf courses and parks.

In 1977 the recycled-water distribution system, which consisted of 14 miles of pipeline, delivered 20 million gallons a day to irrigate non-residential landscaping throughout St. Petersburg. The city's initial concerns about customer acceptance of recycled water were soon forgotten as the repurposed H_2O was put to use throughout the city. The next worry was how to expand the network to include more customers.

By the early 1980s, with increased population and water demand, the St. Petersburg water utility sought to expand its use of recycled water to residential customers. Before the utility could do that, it had to conduct additional research to determine the level of treatment necessary to be safe for residential uses. The utility increased its level of treatment—though it still was not up to drinking-water standards—and gained approval from state and federal authorities for residential irrigation and other non-potable uses. In 1986 St. Petersburg began a $110 million expansion of its recycled water system, and by 1990 the recycled water used in St. Petersburg expanded to include residential use. The recycled distribution system has grown from the original network of 14 miles of pipeline to 291 miles, and from 19 million gallons a day to 37 million gallons a day.

Recycled Water Not Enough for the Thirsty Lawns

St. Petersburg made significant advances in meeting its water demand with its pioneering effort in recycling. Yet the city still could not keep pace with demand. Even with recycled water, St. Petersburg will never have enough of it to satisfy the irrigation of lawns and other water-thirsty landscaping uses. There will always be a water gap between supply and demand because the amount needed for landscaping is greater than the amount available for recycling. According to the St. Petersburg Water Resources Department, the average residential lawn uses about 30,000 gallons per month, while the average household discharges 5,000 gallons a month to the sewer system. As of 2000, St. Petersburg has begun placing restrictions on the amount of reclaimed water that can be used for landscaping, limiting watering of lawns with recycled water to three times a week, and with potable water to once a week. Even with these new restrictions, however, the city of St. Petersburg still needed more water.

Seeking Additional Sources of Water: Desalination

Tampa, St. Petersburg's more populous neighbor across Tampa Bay, is also facing a water crisis. In 1974, the two cities and several surrounding towns had joined forces to form a new regional water authority through which they could decide how to share common water sources and find additional sources of water.

In 2007, through the regional authority, St. Petersburg and Tampa made investments to secure additional sources of water. The regional authority has invested over $158 million in a water desalination plant that produces about 25 million gallons a day. In late 2009, the desalination plant suffered a power outage and temporary shutdown from being pushed to its maximum capacity for too long during a period of drought.

While the desalination plant has provided—when it was functioning—some water to the Tampa Bay water users, it is divided among all the Tampa Bay users and comprises very little of St. Petersburg's water. And so in many ways it is "a drop in the bucket" as St. Petersburg frequently goes into a water deficit with their water-thirsty landscaping.

Need Relief: Cash for Grass?

St. Petersburg residents are being encouraged by the city to replace their lawns with less thirsty plants that are more appropriate for growing in the arid climate of southwestern Florida. So far, however, St. Petersburg's elected officials and the Water Resources Department have yet to make serious advances in convincing the residents to replace their lawns with more Florida-friendly landscaping.

St. Petersburg could take a lesson from several American municipalities. In Las Vegas, San Antonio, El Paso, and, more recently, Los

Angeles, water utilities have paid their customers to remove their water-thirsty lawns.

Las Vegas started the first "cash for grass" program in 2003. There, the rebate for turf removal is $1.50 per square foot for the first 5,000 square feet of lawn; after that the rebate decreases to a dollar per square foot, with a limit of $300,000 paid out to each customer per fiscal year. Several other cities are using such rebates as a means to limit use for water-thirsty lawns. In San Antonio, Texas, the city water utility pays its customers based on overall lot size with a limit of $400 per customer. Los Angeles joined dozens of other California cities and began a "cash for grass" program in 2009. The Los Angeles Department of Water and Power is paying its customers a dollar per square foot. The cash rebates are paid not just for ripping out the lawn; most programs require that participating customers also upgrade their irrigation systems to more efficient methods, such as drip irrigation.

These programs have been very successful in reducing water use. Patricia Mulroy, head of the Southern Nevada Water Authority, had this view of their program: "Although spending more than $100 million to get our customers to buy less of our product would seem counterintuitive, this strategy of investing in water efficiency has paid huge dividends that will continue in perpetuity. Already, enough grass has been removed to lay a roll of sod more than halfway around the earth. As a result of this and other conservation programs, Southern Nevada consumed 26 billion gallons less water in 2009 than it did back in 2002, despite a population increase of 400,000 during that span."

A similar "cash for grass" program could be appropriate for St. Petersburg and Tampa. A comparison of the costs and benefits of expanding its water desalination plant versus the "cash for grass" rebates could weigh in favor of the rebate.

St. Petersburg at the Crossroads

As the year 2009 came to a close, the St. Petersburg area had yet to put in place a program to give its residents an incentive to reduce their habitual lawn-watering. For officials with the St. Petersburg water department, it doesn't seem like this will happen anytime soon. John Riera, the water resources manager for St. Petersburg, considers the move to remove lawns as "running into a political buzz saw." He doesn't see much hope for change even when there have been water shortages: "Everybody forgets when it rains." While others may disapprove of St. Petersburg's failure to wean its customers from their water-thirsty ways, it must be acknowledged that without its pioneering efforts in water recycling, the area would be in dire straits. We are hopeful that we will see a revival of that inventive spirit and some ability to stand up to the "political buzz" as the area comes to grips with its water situation.

EACH OF THE government-run utilities reviewed in this chapter was able to make significant contributions by using existing technologies.

Orange County, California, the nation of Singapore, and St. Petersburg, Florida, were all able to provide greater water security to their customers without adversely affecting their neighbors (that is, they didn't have to take water from someone else) or harming the environment. In fact, the actions of all three utilities meant greater protection for the surrounding natural resources.

It is clear that these success stories can be replicated, and indeed some of them have been. But they have been replicated on only a limited basis. The major drawbacks for some cities or regions might be the cost of implementing a recycling program. But if you look at the case of Orange County, it is clear that the program did prove to be cost-effective, especially compared to the costs of seawater desalination.

Aside from cost, the most common impediment to implementing such programs is the will to act, or the fear of taking a risk. In some cases, *failure* to act might be the greater risk. Thanks to the bold, visionary action of the Singapore water utility, for example, the people there no longer have their water supply at the mercy of political pressure or financial uncertainty.

As wasteful as individual users might be, their wastage is slight compared with that of the agriculture industry, the largest users of water. Reducing the large misuse of water will require more than a few individuals showing leadership and wise management; it may take the collective and persistent action of many individuals and at many levels of government to rein in farmers and ranchers.

CHAPTER THREE

TAMING THE BIG USER

Improving Agricultural Water Use

Around the world, agriculture is the biggest user—and the biggest waster—of water. This one industry uses three times more water than all the other users (municipalities, industry, and commerce) put together. Agriculture consumes 70 percent of all the fresh water that is available on earth. This amount includes natural water supply from rainfall as well as from engineered irrigation. In several agricultural regions of the world, the water supply is already under stress. To protect our food supply for the foreseeable future—for the next four decades—we need to secure enough water for our farms to grow our food and feed our livestock.

In this chapter, we will look at three agricultural success stories, featuring two of the largest farming states in the United States, Nebraska and California, and the third example takes place in Australia's largest farming region, the Murray-Darling River basin.

We begin in Nebraska, where modernization of irrigation methods and agricultural practices put into place by a Great Plains farmer—who considers farming his life—resulted in conserved water and more productivity in his fields.

RISING CITY, NEBRASKA

Eugene Glock, a very youthful seventy-seven-year-old farmer, stands six feet tall and wears a Stetson hat and cowboy boots—the perfect picture of a western rancher. One should not be fooled, however, by his down-home, folksy speech patterns, for here is a man who really understands how to save precious water on the arid high plains of Nebraska. On land farmed by his father and his grandparents before him, Glock and his wife farm 720 acres of mixed crops, mostly soy and corn, using a "center pivot" irrigation system invented, designed, and manufactured in Nebraska. Glock knew that in order to grow crops in parts of the world that do not have adequate and timely rainfall, farmers must

Agriculture is by far the largest user of water in the United States and around the world, typically consuming three times more water than all the other water users (municipalities, industry, and commerce) put together.

be prepared to provide water in enough time to stop the crops from wilting. In order to anticipate such shortages farmers have devised a variety of mechanisms to water only when needed.

From the earliest times in history, farmers have relied on diverting rivers or digging wells to bring water to their croplands. Over time these systems have become more complex and sophisticated, ranging from simple gravity systems and river diversions or dams to deep wells pumping to field channels, drip systems, and ultimately to the big-wheel center-pivot systems so useful and beloved in the arid and semi-arid parts of the United States.

Historically, farmers have had no choice but to pay close attention to the amount of moisture in the soil that enables the roots to develop and the plants to thrive. Various folk methods, such as rubbing the soil between your fingers—to measure moisture—have been used since time immemorial. In arid regions, however, these folk methods are not very reliable, and modern farmers must rely on a variety of electronic sensors buried in the soil at various depths to get a more accurate reading of the water available during crop growth. In addition to the soil moisture measurements, farmers like Glock use monitors to determine the actual amounts of water being absorbed

> Over the past two decades, by clever engineering, Glock has managed to reduce his use of water by more than two-thirds while doubling his crop yields at the same time.

by the crops and infrared devices to measure the temperature at the top of the crop canopy.

These devices, scattered widely over the field, can communicate with a central computer that regulates the rotation speed of the "big wheel" (the center-pivot system) and the amount of water to add to each section of the field; and they determine when a particular piece of land has received enough moisture to keep the growing crops healthy.

The relative ease with which Glock was able to reduce the amount of water applied to his crops and simultaneously increase his crop yields is of fundamental importance to sustaining agriculture and world food supplies into the future. Glock's experiences could help millions of farmers around the world, who are facing similar semiarid conditions, to save water and increase their crop yields. Using these readily available technologies could go a long way toward resolving the global water crisis and could lead to sustainable food production.

> *Even a 10 percent reduction in water applied to irrigation globally could forestall extreme water shortages by many decades.*

With an average annual rainfall of only 24 inches and frequent droughts, Nebraska is a hard place to farm. In his first year, 1955, Glock lost almost his entire crop to drought, so in 1956 he installed a 260-foot-deep well into an artesian aquifer. Artesian aquifers, found in many places around the world, are a natural phenomenon with curious properties: If you drill a well into the aquifer, the water will rise to the surface under its own pressure due to hydrostatic forces, which act on the underground water from the higher elevation of the water recharge area, where rainfall seeps into the ground, often dozens or more miles from the well site. Of course, as more wells take water from the aquifer, the artesian pressure drops accordingly and the water levels in the wells will drop. From the beginning Glock's artesian aquifer never forced water clear to the surface. At the time he first drilled the well, the water rose to about 50 feet below the surface and he had to pump it up to the ground surface. Now, when all of his neighbors' wells are running, his water table drops by another 40 to 50 feet. This lowers the water to the point where, by late July, he is pumping from the bottom of the hole and gets only about 300 gallons per minute instead of his original 1,200 gallons per minute.

Two of Glock's wells drill into the huge Ogallala Aquifer that stretches more than 900 miles southward from Nebraska through Kansas, Oklahoma, and into Texas. Approximately 27 percent of the irrigated land in the United States overlies this aquifer system and pumps water from it. In recent years the aquifer has been severely stressed. Water tables drop in some locations by as much as one to three feet per year due to excessive withdrawals for irrigation. However, in the part of the Ogallala overlain by Nebraska, water tables are now rising because of widespread use of the water-efficient center pivot irrigation systems.

Center pivot irrigation systems are so large and regular (perfect circles of up to a mile in diameter) that they are among the Earth's surface features that are easily seen from the International Space Station. (You may even have seen them when flying over the Great Plains or east of the Rockies en route to California.) They appear as bright green circles on dry parched landscapes in Saudi Arabia and the American West alike. Center pivots, as shown in Figure 3.1, are large wheel-mounted pipes that rotate slowly and cover a perfect circle, with water dripping out of the pipes onto the ground as they rotate. For field crops, the center pivot is much more sparing in its consumption of energy and water. The center pivot is typically composed of one arm a quarter mile in length anchored to a rotating tower at the center of the irrigated plot, which is connected to a pump that draws the water from the well. Hence, the water is applied to the plants not by running the water over the fields in ditches and furrows, or from fixed sprinklers, but by sprinkling the water directly onto the crops from the moving overhead pipes. Prior to the invention of the center pivot technology, sprinklers (like giant lawn sprinklers) were set up at fixed locations in the fields to spray water on the crops. Later, "big guns" were used to fire water great distances over the fields from a fixed location. Both of these methods use large amounts of energy, and, worse, much of the water sprayed is lost to evaporation before it even reaches the crops. The most important contribution of center pivot systems is that they allow for irrigation in undulating or non-level landscapes. Under grav-

Figure 3.1 Photo of center pivots operating in a field in the Snowy Mountains Region of South West Australia. (*Courtesy of Valmont Irrigation*)

ity systems the land has to be carefully leveled, often using laser-guided grading tractors, in order to make sure that the water reaches the plants evenly. The center pivot irrigation system does not require precise land leveling, which is very expensive.

The arm of the center pivot is driven by either hydraulic power of the pumped water or electric motors, and is energy-efficient. We've already mentioned that this system saves water by reducing evaporation and does not require land leveling. The center pivot's main advantage over other technologies, however, is its ability—when coupled to in-field

Figure 3.2   Aerial photo of center pivot irrigation systems near Alamosa, Colorado. (*Courtesy of Valmont Irrigation*)

moisture sensors through a central computer—to irrigate varied crops requiring different water applications in the same field at the same time using the same irrigation equipment. The computer controllers built into these systems enable a wide variety of clever monitoring and sensing techniques that further economize on water use.

Eugene Glock is a user of several advanced soil moisture sensing techniques. For Glock, and every farmer in the world, really, it's essential to understand the details of the factors influencing plant growth.

In traditional agriculture farmers rely on their five senses to assess the amount of water available in the soil that is necessary to prevent wilting and temperature stress on the growing crops. The farmer then decides how much water and where on the farm to irrigate during a ten-day period. Glock (like other farmers in the developed world) reduces the guesswork by using readily available science and technology. For example:

- Glock subscribes to a weekly service that samples moisture in the soil at depths of 6, 12, and 24 inches.
- He uses an evapotranspiration (ET) gauge that measures the amount of soil moisture being used by the plants growing in the field.
- He combines these ET gauges with infrared sensors to measure the heat emitted from the canopy of his crops to schedule the water released through the center pivot to each part of his field in order to avoid heat and moisture stress on his crops.

This information, coupled with the computer-controlled technology of the center pivots, allows him to put different amounts of water in different parts of the land, depending on the actual soil moisture and heat stress at particular locations in the field and the tolerances of the specific crop in that part of the 137-acre field that a typical center pivot can water. While he activates the controls on his irrigation equipment manually, there are neighboring farmers who rely upon remote sensors implanted in their fields to activate the irrigation equipment.

The results of Glock switching to center pivot irrigation have been spectacular. Since he has both dry and irrigated land, he is able to chart the dramatic shifts in crop yields over time. For his two major crops over the past 20 years, soybeans and corn, irrigated soy production at 60 bushels per acre is almost double that of dry land production; and, irrigated corn at 220 bushels per acre is almost three times the dry land production. Since his start in 1956, he has been able to reduce the amounts of water given to the crops per year per acre from 30 inches down to 5 or 6 inches, a saving of more than 80 percent. Not all of the savings are due to the irrigation, however, because Glock also practices no-till agriculture, which saves a few inches of soil moisture. Also, the yields have improved with agronomic research into new varieties, and better fertilizer and herbicide applications.

All of these gadgets—gauges and sensors—are inexpensive and are available to farmers worldwide if they choose to use them.

Soybean yields grew from 45 to 58 bushels per acre and corn yields went from 145 to 189 bushels per acre with over 50 percent less water.

Nonetheless the improvement of yields due to smarter and better irrigation techniques is most impressive. The wide use of these methods in Nebraska enabled the state in 2008 to almost match California's 9.6 million acres of irrigated land.

The Trend Setter

When asked about his influence on neighboring farmers, Glock replied, "As for my influence on better use of water, I wish I could say that everyone looks to me for advice on the subject, but that would be inaccurate. However, some of my neighbors who are also concerned about the future availability of water do watch 'the crazy stuff Glock does.'" He pointed out another factor that has been a recurring theme in water management: "The high cost of fuel a couple of years ago did more to bring about change. It was easier to do it the way they always did, but when it really impacted their profit margins, they changed quickly."

According to Glock, the biggest influence on many farmers comes from the crop consultants they use for advice. These consultants in turn get their information about production, including efficient use of water to maximize profits, from the University of Nebraska–Lincoln Extension Service. The extension services are at the forefront of promoting the use of moisture-sensing equipment, ET (evapotranspiration) gauges, infrared sensing systems, and the devices that shut off pivots when it rains. The consultants also work with seed companies and farmer cooperatives to provide equipment to these entities' customers at a subsidized cost.

Glock's expertise has been widely recognized in Nebraska. He serves on the Water Research Panel at the University of Nebraska at Lincoln (UNL) and has served as adviser to senior federal Department of Agriculture officials. He claims that achieving more efficient methods of irrigating is a slow process, but gradually people are beginning to realize that it is to their benefit as well as that of future generations to use water in the most efficient manner. What we have learned from Eugene Glock's experience demonstrates that inexpensive and readily available technology can be used to improve food production in the poorer regions of the world.

In the next case, we'll look at how a large California farming region was forced to adopt conservation measures by state and federal officials concerned about wasteful water use and the necessity to divide the Colorado River among seven states. The result is that the largest user of the Colorado (and possibly one of the more wasteful ones) is learning new ways to conserve water.

IMPERIAL VALLEY: IT'S ALL ABOUT THE WATER

Imperial Valley, an area of one million acres in the southeastern corner of California east of San Diego, is also one of the state's largest agricultural

regions. You can thank (or blame) the
Imperial Valley for the broccoli served at
a Christmas dinner. The Valley's farmers claim to grow two-thirds of the nation's winter vegetables. The Valley, with
its 150,000 residents, is the largest user

> *Imperial Valley, home to 150,000 residents, provides two-thirds of the United States' winter vegetables.*

of the Colorado River, which forms its eastern border. Every year, Imperial Valley uses 3 million acre-feet of water, which would be enough for
the 8 million residents of New York City
for three years. (How much water is an
acre-foot? It is enough to cover one acre
of land with one foot of water. A more
descriptive answer is that one acre-foot is
325,800 gallons of water. Two acre-feet
would fill an Olympic-sized swimming
pool.) Now, with strong incentives from
state and federal government, the Imperial Valley is learning to use water more wisely.

> *Every year, Imperial Valley uses 3 million acre-feet of water; enough water to last the 8 million residents of New York City for three years.*

A Dry Wasteland Becomes Imperial

The Imperial Valley's history is all about water. "Without water the
valley don't exist," said Mike King of Imperial Irrigation District, the
government agency that manages the Valley's water. To understand

fully this connection to water, it is important to know how a dry and barren desert became the "Imperial Valley."

In the mid-sixteenth century, a Franciscan friar traveling with Spanish explorers came upon the dry, dusty valley with its scorching temperatures; he referred to it as a "deadly place." Two hundred years later, John Audubon, the son of the famous naturalist John James Audubon, upon visiting the valley, called it "most melancholy." The Imperial Valley, then known as the Colorado Desert, was indeed dry, dusty, and deadly, with summer temperatures consistently above 100 degrees. To transport goods from Pacific ports to Arizona, it was necessary to traverse a sweltering wasteland the size of Rhode Island.

The Valley's agricultural potential was first identified in 1853 by William Blake, a geologist from Yale. Blake was part of a surveying party trying to find railroad routes eastward from the Pacific. As he passed through the desert, he noticed an ancient shoreline at the foot of the mountains, which supported the notion that there once had been an inland sea in the Valley.

Blake's analysis of the land also showed that the soil would be a fertile land for crops, if it were ever irrigated. He concluded that this dry valley could be irrigated by a canal diverting water from the nearby Colorado River.

Decades after Blake's discovery, there were several efforts to bring the water from the river to the Valley. In 1901, one of the first attempts was made by the California Development Company, which coined the

Figure 3.3   This 1904 photo shows the Colorado Desert before it was irrigated by the Colorado River and became the Imperial Valley, which supplies vegetables throughout the United States. The photo illustrates the desert conditions and provides a glimpse of the shoreline of an ancient sea that once existed in the Colorado Desert. (*U.S. Geological Survey*)

term "Imperial Valley." The California Development Company engineered a "cut," or opening, into the banks of the Colorado River and tried to direct the water westward through a canal to the Imperial Valley. The company did not have the U.S. government's authorization to take water from the Colorado River, so their "cut" into the Colorado was made just below the Mexican border. These first attempts, which delivered some water to the Valley, were largely unsuccessful because the water was blocked by a buildup of silt in the canal.

The Desert Becomes a Verdant Valley

In 1905, the company made another cut into the Colorado and water began moving into the Valley. But this time the company was *too* successful. With the winter came floods, and the breach became a chasm 150 feet wide; by August of that year the Colorado River began to flood the Valley. With the winter rains of 1906, the cut exploded open to a width of 2,700 feet and a depth of 40 feet. Because the Valley is below sea level, the water was drawn into the lowest elevations, flowing south to north toward the deepest end of the Valley. A fierce battle was on between the raging waters of the Colorado River and the company's effort to establish controlled irrigation of the Valley.

The company's uncontrolled irrigation of the Valley, though decades before Walt Disney's animated film *Fantasia*, seemed to be a forerunner to Mickey Mouse's role as the sorcerer's apprentice. The company, like Mickey, had dreams beyond its capacity to manage the water. Because they did not have the resources to control the flooding river, the company requested help from the U.S. government. The company's requests were rejected, however, because the cut in the riverbank had been made in Mexican territory.

The task of controlling the deluge was taken up by the Southern Pacific Company, a railroad with tracks stretching across the Valley. With its route at risk, Southern Pacific had a strong interest in taming the raging waters. The railroad stepped up to the challenge, spending the next ten months building bridges over the wide gap in the Colorado River bank. It then brought in trainloads of boulders and dumped them into the opening in the bank of the river. By early February 1907, the breach in the Colorado River was contained.

By this time, the formerly dry and dusty Colorado Desert had become the irrigated and fertile Imperial Valley. As geologist William Blake had predicted 50 years earlier, the Valley was a great place for crops—all it needed was water. The formerly dry Salton Sink, at the upper (northern) end of the valley, became the inland Salton Sea, covering 400 square miles with a depth of 80 feet. It became California's largest body of inland water.

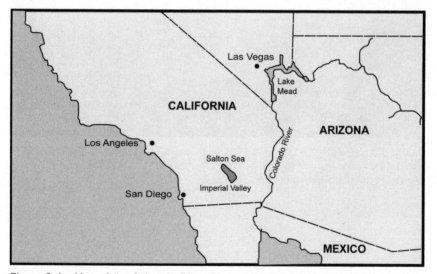

Figure 3.4   Map of the Colorado River Basin, showing Mexico and three of the seven states that draw their water from the Colorado River. The Imperial Valley draws more water than Mexico or any other state—more than three million acre-feet per year.

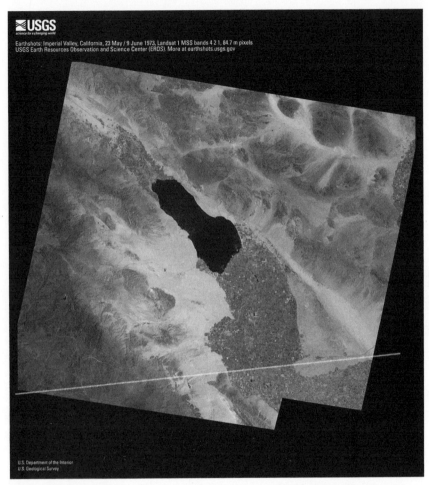

Figure 3.5   Satellite photo of Imperial Valley and the Salton Sea, showing the agricultural development throughout the 500,000 acres of the valley. The Mexican border is marked by dotted white line. (*U.S. Geological Survey*)

The Imperial began to take its place as one of the nation's most productive agricultural valleys. Over its irrigated 500,000 acres, farms produce large quantities of an almost endless variety of crops. With plentiful water, fertile soil, and constant sunshine, this desert became a lush agricultural valley growing a wide variety of crops all year long.

A Few Hundred Farmers Control the Valley

Imperial Valley's vast agricultural economy is controlled by a relatively small number of people, the farmers. Most farms are family-owned and very large. The average size of an Imperial farm is 950 acres; the average in the rest of California is about 230 acres. Approximately 500,000 acres of farmland is controlled by 450 farming entities.

The farmers controlling Imperial Valley's economy have had little or no motivation to use water more efficiently. They were getting plenty of water at a very cheap price; Imperial's farmers pay a fraction of what their counterparts pay in neighboring counties. As a result, the farmers have had little concern about their irrigation methods. In the Valley, irrigation is accomplished by dispersing water through large canals and then to channels that wind through the fields. Many of those channels are unlined dirt troughs. As a result, excess water—beyond what is needed for the crops—is left to seep into the ground or evaporate into the hot Valley air.

Imperial Valley Wastes Enough Water to
Support New York City

The Imperial Valley farmers had long used as much water as they
wanted and were accustomed to taking more than they needed. For
decades, it appeared as though these practices would continue indefi-
nitely, but a ruling by the State of California began to change "business
as usual" in the Imperial Valley.

In 1984, the state's water board documented Imperial Valley's excessive water use. The board found that out of the 3 million acre-feet
of water going into Imperial Valley, almost 1 million acre-feet—or a
one-year supply for 8 million urban dwellers—was lost and not used for
crop production. The board also determined that it was reasonable for
the Valley to be able to conserve 200,000 to 400,000 acre-feet, or at
least 10 percent, of the water it used. Based on its findings, the state's
water board declared the Valley in violation of California's constitu-
tion, referring to the provisions requiring that water be put to "bene-
ficial use" and that "water waste or unreasonable use . . . be prevented."
Imperial Valley challenged the ruling, but it was reaffirmed by both
the state's water board and the California courts.

Soon after the state's ruling, Imperial began negotiations to transfer water to the Metropolitan Water District (MWD), a very large
water district that provides water to over 18 million people in several
Southern California counties, including Los Angeles, Orange, and San

Diego. The Imperial Irrigation District (IID) and MWD began negotiations in 1984 and finalized in 1989 an agreement to transfer 100,000 acre-feet of water per year, for a period of 35 years. In return, MWD would pay Imperial $222 million—over the duration of the agreement—for conservation projects.

While the agreement with MWD appeared to give Imperial Valley some breathing room, within a few years, the pressure to further reduce its water use intensified. The U.S. Department of the Interior began to move ahead with its plan to limit California's draw from the Colorado River to the amount decreed by a 1928 federal law, a "firm allocation" of 4.4 million acre-feet of water per year. For years, California had been taking closer to 5.2 million acre-feet per year (with the largest portion to Imperial Valley). With pressure from the State of California and the Department of the Interior, Imperial Valley was concerned about being forced to give up more of its water.

With this backdrop, in the mid-1990s Mike Clinton, then head of the Imperial Irrigation District, began discussions of transferring significant amounts of water to San Diego. He wanted Imperial Valley to control its own destiny. He saw the state and federal governments giving the Valley one of two choices: arrange a water transfer agreement with San Diego and use San Diego's money to improve the Valley's efficiency, or let the other states and urban counties take the Valley's water by political muscle or legal mandate.

Under this proposed transfer, San Diego would receive 200,000 acre-feet of water from Imperial, and Imperial would receive at least $50 million a year for a term of 45 years, with an option for another 30 years. The money from San Diego would allow Imperial Valley to upgrade its irrigation systems and operate more efficiently. In return, the state and federal governments would allow a 15-year transition for Imperial Valley to reduce its allotment from the Colorado River. The water saved through conservation would be transferred to San Diego, giving that county a reliable source of water.

Imperial Valley's "Shotgun Wedding" with San Diego County

In 1996, when the Imperial Irrigation District offered to sell the Valley's water to the San Diego County Water Authority, San Diego was happy to buy. In the words of Maureen Stapleton, the head of San Diego's Water Authority: "We were coming off a devastating drought; we jumped at the chance to diversify our water supply." The *New York Times* characterized the water transfer as a chance for San Diego County to buy its freedom.

While the management of Imperial Irrigation District wanted to transfer water and San Diego wanted to accept the water, many of the Valley's farmers were opposed to the idea. One Valley resident referred to the water transfer agreement as a "shotgun wedding." From 1995 through 2002, the negotiations between San Diego and Imperial took many turns back and forth from a "done deal" to "dead on arrival." Within the competing interests for the Colorado River water, there were several agendas.

In support of the agreement were the federal government and all of the states dependent upon the Colorado River. The federal government was concerned with balancing the needs among the seven states that draw from the Colorado River. For the states that had fought bitterly over the water, the transfer agreement between Imperial and San Diego became a linchpin to reach some accord on the use of the Colorado. In 1998, then Secretary of the Interior Bruce Babbitt warned that without the water transfer agreement there would be a "very bleak future for California and the Colorado River."

Environmental groups supported the water transfers. Over the past several decades, the Colorado River had been over-allocated. The 1,400-mile river that once flowed to the Sea of Cortez was being diverted to provide water to agriculture and 20 million people. Further reduced by drought, it was not much more than a trickle by the time it reached the sea. Though they had little hope of ever restoring the river to its former glory, the environmental community was supportive of conservation measures in Imperial Valley.

Environmentalists were also concerned that the water transfer could have a negative effect on the Salton Sea. Scientific analysis had confirmed that decreased run-off water from the farms would increase the sea's salinity, negatively impacting the habitats of migratory birds and other area wildlife. Conservation and water transfer to San Diego would result in less farm run-off, so there needed to be a plan to mitigate the impact on the sea. The proposed mitigation plan was that, for the first 15 years of the agreement, there would be fallowing of farmland (pulling farmland out of production) to be gradually replaced by conservation over the 15-year period. During this transition period, state and federal agencies were to design and implement a plan to maintain the sea.

As expected, most of the Valley's farmers opposed the water transfer. They disliked the demand to conserve water. The additional requirement of compulsory fallowing portended economic devastation and made the proposed transfer unacceptable to them. The farmers were accustomed to getting plenty of water at a cheap price; the requirements and limitations seemed over the top.

Caught in the Middle: The Farm Workers

While the farmers were concerned with maintaining plentiful and cheap water, the farm workers were concerned about their ability to scratch out a meager existence. Despite the agricultural abundance of the land, Imperial Valley is not the "land of milk and honey" for the farm workers. From its early days as an agricultural valley to the present, it is a tough place for the people in the fields.

In the 1930s Imperial Valley's farm workers were the subject of Dorothea Lange's photography chronicling poverty and misery, including her famous picture, "Migrant Mother," one of the iconic images of the Great Depression, photographed in nearby Nipomo, California. In early 1936, Lange described the Imperial Valley: "I moved from Nipomo to the Imperial Valley because of the conditions there. They have always been notoriously bad as you know and what goes on in the Imperial is beyond belief. . . . [I]n the meanwhile, because of the warm, no rain climate and possibilities for work the region is swamped with homeless moving families."

Decades after Lange's photographic study, the Imperial Valley continues to offer a meager existence to the majority of its agricultural workers. 2007 data confirmed that the Valley's residents have poorer economic opportunities than the state average. The median household income in the Valley is $34,000, while the statewide average is $60,000. There are similar disparities in high school graduation rates.

For the farm workers, taking farmland out of production would result in unemployment and an even bleaker existence in the Valley. Advocates for these workers lobbied for an economic relief plan to compensate for lost jobs.

High Stakes Game in Vegas

Negotiation of the transfer agreement continued in earnest for several years. The pressure on Imperial intensified with threats by then–Secretary of the Interior Gale Norton to cut back on California's water allotments if the transfer agreement was not approved by the end of

2002. In December 2002, the elected board of the Imperial Irrigation District turned down the agreement by a 3–2 vote. Secretary Norton responded by turning up the heat. In late December 2002 in a speech delivered in Las Vegas, Secretary Norton likened her battle with Imperial Valley to a high-stakes poker game:

> If this were a game, it would be time to lay our cards on the table and demonstrate that the Department of the Interior has not been bluffing. However, this is not a game, it is serious business. But I will lay our cards on the table here this morning.

Indeed, she was not bluffing. On January 1, 2003, Norton slashed California's water allotment from the Colorado River and forced the Imperial Valley and its farming interests to make some tough decisions. They tried to restore the water allotment through the courts, but were unsuccessful.

Deal Consummated, Imperial Valley Agrees to Conserve Water and Transfer Water to San Diego

Finally, desperate for water, in October 2003 the board of the Imperial Irrigation District approved, by a 3–2 vote, the water transfer agreement. Stella Mendoza, a board member, voted against the agreement and described the transfer as "farmers farming water instead of farming crops." Bruce Kuhn, a board member who voted for the transfer, saw it this way: "There are too many people that agree that this is the thing to do. For the first time in 50 years, we have the federal government, the state of California and six Colorado River Basin States all in agreement."

This water transfer agreement was the largest of its kind in the nation and contained these key provisions:

- San Diego County Water Authority commits to pay billions of dollars to the Imperial Irrigation District to improve the efficiency of Imperial Valley's agricultural output (starting at $258 per acre-foot in 2003 with provisions for increases for inflation) over 75 years (the initial term being 45 years with an option for another 30 years).
- Imperial Valley will provide San Diego County with an additional 200,000 acre-feet of water each year. This is a significant amount of water for San Diego County, as officials project an annual demand of 813,000 acre-feet by 2020, up from its current demand of 600,000 acre-feet for its growing population.
- Through the first 15 years of the water transfer agreement, land fallowing is the chief means of generating conserved water, mainly to lessen the environmental impacts upon the Salton Sea.

- A $50 million fund will be established to assist farm workers affected by the fallowing.
- As a result of this landmark agreement, Imperial Valley protects itself against challenges, for the foreseeable future, to claims of its "unreasonable" use of water.

The early results of the transfer agreement are quite positive. The Imperial Irrigation District has made a number of improvements in its irrigation infrastructure, and many farmers have implemented the fallowing program. The Imperial Irrigation District, which manages 1,438 miles of water canals that funnel water to farms throughout the Valley, has begun implementation of a series of upgrades to those canals that will conserve close to 75,000 acre-feet of water. These upgrades will include better measurement and control of water being delivered to the farms to reduce the amount of spillage or waste. While some farmers and elected officials in Imperial Valley continue to challenge the validity of the transfer agreement, the Irrigation District continues its efforts to conserve water. Imperial Valley exceeded its water-saving goals in 2009 and is projected to be on track for 2010.

The urban users in San Diego will receive 200,000 additional acre-feet of water and in return will give billions to improve the water conservation infrastructure in Imperial Valley.

Even Stella Mendoza, the board member who had opposed the agreement, supports the conservation and efficiency effort. Mendoza observed, "Hey, it's a legally binding agreement and we must honor it. We can use water better; the technology is available for us to improve our water use."

Compared to other agricultural areas of California, Imperial's efforts at improving water efficiency are far from the leading edge. The prices Imperial pays for abundant water are similarly out of step: Imperial farmers pay $17 per acre-foot for water, while farmers in neighboring San Diego County pay over $450. Still, it is a hopeful sign that Imperial Irrigation District describes the agreement in these terms: "[It] provides the methods and the means to allow IID to elevate its Colorado water use to efficient twenty-first-century standards and ensure the continued availability of this precious supply." If they stay with their plan, Imperial Valley might begin to use water in a way that reflects how truly dependent their livelihood is upon it.

The agreement is an opportunity with great potential to improve the efficiency of one of the largest agricultural regions in th country. From our viewpoint, if the Imperial Valley can conserve waer, anyone can. The transfer agreement also shows what can happen when elected and appointed government officials step up and demonrate leadership. Without the pressure by the Department of the interior under both the Clinton and Bush administrations and the suport of elected officials in California, this agreement would never lve been signed;

there would have been no incentive for Imperial to change its behavior. This should serve as a model for other parts of the world, both farmers and public officials.

In our first case of "taming the big user," we have seen how farmers such as Eugene Glock can improve their crops' productivity and conserve water at the same time, influencing neighboring farmers to do the same. Next we'll look at yet another method to improve water use efficiency and agricultural productivity, this time in Australia.

IN AUSTRALIA, MORE PRODUCTIVE FARMING
IN THE MURRAY-DARLING BASIN

Australia is the second driest continent in the world. Its largest agricultural producing area, the Murray-Darling River Basin, is responsible for 40 percent of the country's agricultural income. Since 2001, however, it has been in the grip of a severe drought. Yet, despite the long and devastating drought, the Murray-Darling Basin has stabilized agricultural production through an innovative system that allows farmers to trade their water rights, that is, to buy and sell their property interests in water. Once they learned to recognize and monetize the value of water, the farmers began to use it more efficiently.

The Murray-Darling River Basin, Australia's largest river basin, is formed by the Murray River and its main tributary, the Darling River. The basin covers an area of over 385,000 square miles—about the size

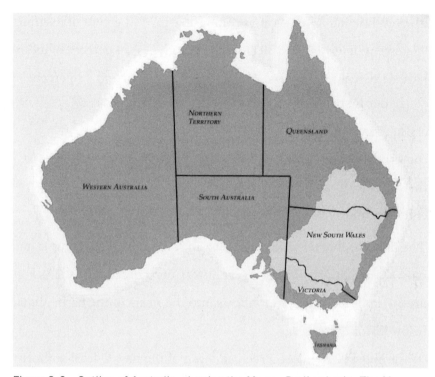

Figure 3.6   Outline of Australia, showing the Murray-Darling basin. The Murray-Darling Basin is Australia's largest agricultural region and is the size of Spain and France combined.

of France and Spain combined, and similar in size to the Colorado River basin—and spans across four states: New South Wales, Victoria, Queensland, and South Australia. The basin includes only a small part of South Australia, yet it provides close to 90 percent of the public drinking water to that state's city of Adelaide (population 1.2 million). While producing 40 percent of Australia's agricultural income, the

Murray-Darling River Basin is responsible for 53 percent of its grain, 95 percent of its oranges, 80 percent of its grapes, 54 percent of its apples, 62 percent of its pigs, 45 percent of its sheep and 28 percent of its beef.

Drought and Over-Allocation Hit the Murray-Darling Basin

The drought that has afflicted the basin since 2001 is the longest and most severe in the basin's 117-year history of recorded rainfall. As Figure 3.7 indicates, the water inflows into the rivers of the basin are far below historical patterns.

In August 2009, Rob Freeman, head of the new federal authority monitoring the basin, made this forecast: "With another El Niño event predicted to bring dry conditions, the overall outlook for 2009–2010 water years unfortunately remains very poor." To begin recovery, the basin will need several years of above-average rainfall.

The drought has had a severe impact on the farmers' water supplies; in some areas, the supplies have been reduced by more than half. Compounding the effects of the drought, the basin's waters have been over-allocated. For decades, state water officials had allowed overuse of the rivers making up the drainage basin as there was limited federal oversight. Over the years, the river's upstream users have built dams and canals without taking into account the downstream users. The

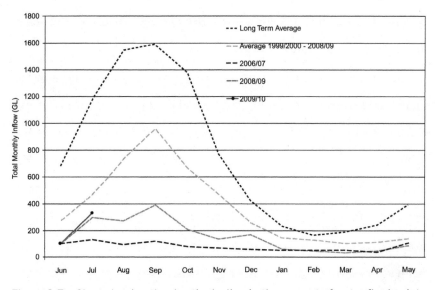

Figure 3.7 Chart showing the drastic decline in the amount of water flowing into the Murray-Darling Basin caused by the multiple-year drought (1999–2009).

farms in the upstream states, which include water-intensive cotton and rice farming, account for 83 percent of the basin's water consumption. The drought and upstream diversions meant less fresh water flowing downstream to the mouth of the Murray River. Lakes located near the mouth of the river that rely on downstream flows experienced a striking increase in salinity levels. In general, the upstream states were allowing their farmers to overuse water.

After years of inequitable use, the Murray-Darling Basin Commission was established to institute some changes in the management of the river basin. In its effort at reform, the commission encouraged progressive water management techniques, notably the water-trading system. This trading system quickly became the darling of the water management community worldwide.

Adding to the Murray-Darling Basin's woes, farmers and governmental authorities were caught flat-footed by the current drought. According to hydrologists and climatologists, the historic drought has been caused in part by climate change, which has resulted in a southward shift in the rain-bearing clouds. While many might argue that an early manifestation of climate change caught them by surprise, they should have anticipated the potential for water shortages in a basin that has had a well-documented history of brutal droughts.

A Positive Note in the Murray-Darling Basin: Water Trading

While the drought and water over-allocation have cast a shadow over the Murray-Darling Basin, the trading system has provided hope for many of the basin's farmers. Under the trading mechanism, inaugurated in the early 1980s and initially operating at a very slow pace, farmers are allotted a certain amount of water that varies according to the sources of water and the size and location of the land. If a farmer needs more than his allotment, the Australian scheme allows him to buy it from another farmer. The system began to accelerate in 1991 when the federal government promoted efficiency and competition in all business sectors.

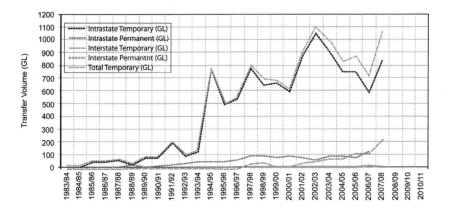

Figure 3.8   Chart shows how water trading in the Murray-Darling basin accelerated in the early 1990s after the Australian government began a program to promote efficiency and competition. This program of selling water allotments between farmers allows the region's farmers to maintain outputs and produce higher-value crops.

The pressure to switch to higher-value crops also grew as the effects of the severe drought increased. In spite of the drought, the farm's output and value remained stable because the trading system allowed farmers to switch to more valuable and less thirsty crops. For example, rice, which uses 1.3 million gallons of water per acre, is a very thirsty crop, whereas grapes use only 400,000 gallons per acre and are worth much more than rice. Cotton, another former staple of the basin, uses 800,000 gallons of water per acre, compared with most fruit at 530,000 gallons per acre.

John Briscoe, Professor of Environmental Engineering at Harvard University, gives a perspective on the trading scheme. "It provides

> *In spite of the severe drought, water trading allowed farmers to maintain their farms' output and value.*

more value per drop of water. It was crazy for rice to be grown in the desert. But, if growing grapes and almonds creates more jobs and better exports, then it is a better use of a scarce resource."

The other noteworthy result of water trading (and another effect of the drought) is that farmers are instituting more efficient methods of using water. In the vineyards, the farmers are switching to more efficient drip irrigation methods and away from the inefficient open furrows directing water to the crops.

Even in the rice fields, the farmers have learned to trim back their water use. The basin's rice farmers claim to have reduced water use by 60 percent over the last ten years: They planted varieties of rice that have a shorter growing season so they will use less water; and they reduced the time allowed for flooding the rice fields from throughout the crop's full maturity to stopping the flooding just prior to its maturity.

Some critics of the trading system claim that the states were allowing farmers to sell water that was already over-allocated. In essence, the states allowed the farmers to sell water when there wasn't any left to sell. Currently, Australia is trying to sort out the best means of restoring the rivers of the Murray-Darling Basin. The federal government and the states have come together in an effort to restore the entire basin and to determine what is best for the common good as opposed to each state taking care of its own interests.

The good news is that the trading mechanism, although not perfect, remains a key component of Australia's water management plans. It has been an invaluable tool in maintaining agricultural productivity during a long and severe drought.

THREE CASES, THREE IMPROVEMENTS IN WATER USE

In each of these cases, technology was available to improve water efficiency and crop productivity, but farmers did not use these tools until they were provided an incentive. For many farmers, there is no motivation to improve unless there is either an economic incentive (Murray-Darling and Nebraska) or an economic incentive and regulatory mandate (Imperial). The encouraging news is that in three farming regions of the world, the largest water consumers were able to significantly reduce their water consumption. In fact, even a 10 percent improvement in agricultural water use saves vast amounts of water, freeing up more water than is currently used by all the cities and industries across the globe.

These improvements in water efficiency and crop productivity are essential for future food production. In 2007, the International Water Management Institute, a nonprofit organization sponsored by

60 governments and private foundations, conducted a study to assess the global needs for agriculture from 2007 until 2050. The Institute's study makes clear that our current agricultural system must make immediate improvements if the world's food growers are to meet foreseeable needs between now and the year 2050. Their findings concluded that:

- Globally, there are sufficient land and water resources to produce food for a growing population over the next 50 years.
- It is probable, however, that today's food production and environmental trends, if continued, will lead to crises in many parts of the world.
- Only if we act to improve agricultural water use will we meet the acute freshwater challenge facing humankind over the coming 50 years.

The International Water Management Institute's findings, taken together with the success reported in this chapter's three cases, give us some hope for our water future.

While we are optimistic that a 10 percent efficiency improvement would be sufficient for the foreseeable future, the impact of climatic change means that 10 percent may not be enough. In the Murray-Darling basin, several research studies indicate that at least part of the drought results from climate change. As we mentioned earlier, climate

change is the wild card in any assessment of freshwater availability. The
trends are troubling. Generally, we
would expect more rainfall, but it is
possible that there will be more extreme weather patterns and changes in
regional precipitation distribution.
Specifically, India and China will expe-

> *A 10 percent improvement
> in agricultural water use
> would free up more water
> than is currently used by all
> the cities and industries
> across the globe.*

rience increased precipitation in the form of more rain and less snow.
Africa and Latin America will face a decline in overall precipitation.
So, while we believe that the lessons learned in these cases are worthwhile, considering the hardships likely to be brought about by climate
change, they may well be the bare minimum of what we must achieve.
It is imperative that the increases in water efficiency and agricultural
production illustrated in our case studies be widely promoted and accepted throughout the world.

CHAPTER FOUR

WANTED:

PUBLIC INVOLVEMENT

THE IMPORTANCE OF INFORMING
AND INVOLVING THE CONSUMER

The successes described in Chapter 2 can be only partially attributed to technological innovation. In large part, these projects were successful because they were able to gain public support. Getting the public's acceptance or financial support is often "easier said than done" because most people take water and wastewater systems for granted and are woefully uninformed.

The following success stories demonstrate that a well-informed and involved water consumer is essential to maintain infrastructure and protect public health. These case studies will show how important it is for a water utility to educate and involve its customers. As for consumers, it is our responsibility to become knowledgeable about our water and wastewater systems that we rely on every minute of every day. The consumer needs to get off of the sidelines and into the game. After all, whose water is it?

SEWER SYSTEMS: OUT OF MIND, OUT OF SIGHT, AND OUT OF MONEY

Across the United States, there are out-of-date, poorly maintained water and sewer systems that continually malfunction, sending untreated feces and industrial waste flowing into our lakes, rivers, and oceans, polluting our drinking water sources and harming our environment. To protect against public health hazards and environmental degradation, massive investment is needed to maintain or upgrade our aging sewer systems; yet, the federal government estimates that our investment in sewer infrastructure falls short, annually, by more than $20 billion.

The difficulty is that most taxpayers won't support investments in public services unless they feel directly affected by their problems. Whereas bridges, roads, schools, and hospitals regularly pique the interest of consumers, it is clear that most folks would rather not think about sewage; meanwhile, ancient and malfunctioning sewage pipes rumble beneath our feet.

Dozens of families in Milwaukee, Wisconsin, could tell you how important it is to have a sewer system that functions properly. Based on data collected from 2002 to 2004, the American Academy of Pediatrics reported that city hospitals experienced a significant increase in emergency-room visits to treat diarrhea after partially treated sewage flowed into Lake Michigan. The untreated sewage went into their waterways because their sewage system was not big enough to treat all the sewage.

Sewage system malfunctions are a problem all over the country, not just in the Great Lakes Region. In 2007, U.S. Senator Frank Lautenberg from New Jersey introduced legislation called the Sewage Overflow Right-To-Know Act. In support of his legislation, the Senator's press announcement cited these statistics: "According to the EPA, between 1.8 and 3.5 million Americans become ill every year from recreational contact with waters contaminated by sanitary sewer overflows."

It is clear that no community is exempt; even in affluent and well-developed communities, constant vigilance is needed to ensure that we do not poison ourselves or our environment. In our first case study, we illustrate one city's efforts to educate its populace about the advantages of a clean, efficient, and sustainable sewage system, and the dangers of the alternative.

THE CITY UNDER THE CITY

San Francisco, California, called "The City by the Bay" because it sits on a peninsula that separates the San Francisco Bay from the Pacific

Ocean, is known for its hills, picturesque cable cars, and panoramic views. The city's charms have long made it an international tourist destination.

But hidden beneath those hills is a severely out-of-date sewer system. Hundreds of miles of pipes pump sewage underneath the city's 49 square miles, and 70 percent of the pipes are over 70 years old, many way beyond the average lifespan. Those pipes carry 80 percent of the city's sewage to the city's Southeast Treatment Plant, a sewage treatment plant that dates back to the 1950s. This antiquated plant belches foul odors into the surrounding neighborhood and is also structurally vulnerable to earthquakes, something that should resonate with San Franciscans, who never know when the next "big one" will hit.

For decades, however, the old treatment plant and the aging sewer pipes were definitely "out of sight, out of mind" for most San Franciscans. In 1998, San Francisco residents showed their lack of awareness of their sewer system when, in a fit of pique, they voted to freeze the sewer rates for eight years. The freeze meant that there was barely enough money in the government-run water utility's budget to perform basic maintenance repairs, let alone embark on the needed upgrades.

Voters had been swayed by claims that the utility responsible for the city's water and wastewater systems was not performing sewer repairs because its money was being funneled from the utility's budget to pay for other city programs. But instead of demanding accountability from their politicians, the voters—clueless about the degradation of the system—approved the freeze. Politicians were also to blame: some supported the ballot measure (freezing sewer rates) in spite of the knowledge that the city's sewer system was in desperate need of an upgrade, opting instead to seize upon the popular issue of keeping sewer rates down.

In 2002, the elected city leaders and the voters had an opportunity to "make things right" with their struggling sewer system. To counter the rate freeze and begin to make system-wide improvements, a ballot measure was prepared that would provide the utility with $1.6 billion to upgrade its drinking water system, and an additional $1 billion to upgrade the city's sewer system. But, in the hours before the measure was placed on the ballot, the city's elected officials and their political strategists, believing that voters didn't care about the sewer system and that keeping the $1 million sewer provision in the measure would endanger the passage of the entire water bill, opted to remove it from the ballot measure. They were probably right. In November 2002, San Francisco voters passed the measure, approving $1.6 million in funding to revamp and rebuild the drinking water system.

But by 2004, the problems with the wastewater system and treatment plant were getting harder to ignore. Besides being dilapidated and dangerous, San Francisco's treatment plant had become a symbol of environmental injustice. Located in San Francisco's Bayview District, one of the city's poorest areas, the plant often draped the neighborhood in a foul stench. The district affected by the sewer plant had just elected a new representative at City Hall. A new member of San Francisco's Board of Supervisors, Sophie Maxwell, lived a few blocks from the sewage plant and knew firsthand about its odiferous qualities. Supervisor Maxwell made a retrofit of the sewage plant one of her top priorities.

There had also been a change in leadership at the water and waste-water utility. The new General Manager, Susan Leal, coauthor of this book, had been on the job only a few months when she was called to an emergency at the sewage treatment plant. During repairs, an aging 30-foot tank containing sludge (sewage from which water has been removed) and filled with methane gas had blown open, causing an explosion that could be heard throughout the neighborhood. Large metal rivets from the top of the tank flew off and landed several city blocks away. The incident helped focus the public's attention on the problems underground. Responding to questions from the press, Susan Leal said: "We can no longer have our sewer system in a precarious state. To me, it's a no-brainer. Without a working water system and without a working sewer system, you don't have a city."

But Maxwell and Leal's concerns did not automatically translate into voter support for an expensive system upgrade. By and large, San Franciscans were still more interested in keeping the rates for city services down. Those rickety pipes underground remained dangerously "out of mind."

Educating and Engaging City Residents

It quickly became clear that unless voters understood the consequences of continuing to defer repairs and recognized the benefits of having a well-run and retrofitted sewage system, nothing would be done. Educating the city's residents and changing their perception of the sewage system became a task for the utility's communications team. They knew they had a tough job ahead of them; their education campaign had to be persuasive enough to result in voter support for significant rate increases. The status quo was no longer an option. Without this support, the old pipes and the treatment plant would inch closer to total malfunction. "The City sewer system had been the political piñata for far too long. We needed to let them know that their sewer system was working for them 24/7," said Tyrone Jue, the head of the communications team. The team set to work developing a multipronged communications plan that they hoped would become a model for engaging community support.

It would be the first citywide outreach program about the city's sewage system. The plan called for $500,000 to be spent over ten months to educate the city's residents. The utility's general manager recoiled at the price tag, but the communications team convinced her that it would not be easy to educate the public and change their misperceptions. The problem was too pressing to ignore. She gave the go-ahead to take the plan to the utility's governing board (the San Francisco Public Utilities Commission) for approval.

One commissioner was strongly opposed to "spending money on PR" and there was some initial hesitation on the part of the other commissioners. But in the end, the majority realized it was important to educate the community if there was going to be any chance of getting support for financing repairs and system upgrades.

Seventy-Five Percent of City Residents Were Unfamiliar with Sewer System

The team's first step was to find out what San Franciscans thought about their sewage system. This was one of the most expensive components of the plan, but the team believed that if they didn't get a true baseline of voter sentiment, it would be difficult to plan effective outreach programs or measures of success. With assistance from outside consultants, they polled more than 800 households. The goal of the surveys was to find out what people knew about the system and how they wanted to receive future communications from the utility. The survey results were informative. The team learned that:

- only 26 percent of respondents were familiar with their sewer system;
- yet 87 percent thought it was important to have a well-run sewage system to protect public health;

- 84 percent thought it was very important to have a well-run sewage system to protect local waters; and
- 93 percent thought it was important that the public be kept informed of the ongoing progress of any repairs or system upgrades.

In an effort to delve deeper into the ratepayers' knowledge and concerns, the team conducted a dozen different focus groups.

Armed with this baseline information, the team created a mailer in English, Spanish, and Chinese that included a detachable, prepaid response postcard and sent it citywide to every household (approximately 300,000). Entitled "City under the City," the mailer provided some basic information about the sewage system, cited the specific problem areas, such as aging pipes, flooding, sewer overflows, and odors, and made the case for the need to retrofit it. People were also asked to prioritize these problems, and record their responses on the card.

The response to the mailer was better than anyone had expected—9,000 San Francisco residents returned their postcards, and 7,000 of them filled in their contact information so they could receive updates from the utility communications team. Many of the returned cards also included additional comments, such as "Fix-it" or "It's important for the environment."

The stacks of postcards gave the team hope that their message was beginning to reach city residents. Comments submitted were used in crafting future messages. The 7,000 respondents that had given their contact information were used as the team's public outreach base for the duration of the project.

San Francisco's Sewers: No Longer Out of Sight, Out of Mind

Over the next eighteen months, the team engaged in a communications campaign to raise public awareness. Some of the education and outreach efforts included:

- Partnering with other public agencies, such as public schools and libraries, to help spread the message. The utility offered classroom tours of the sewer treatment plant and set up kiosks in libraries throughout the city from which to distribute flyers and other information.
- Meeting hundreds of city residents through three citywide public forums, and 60 smaller public meetings held in different neighborhoods across the city.
- Disseminating the utility's message through provocative billboards and ads on the radio and the Internet.
- Sending a quarterly newsletter to every household, updating the public on proposed sewer projects and retrofit planning.

Figure 4.1 Billboard ad produced by the San Francisco Public Utilities Commission to attract attention and gain support for repairs and upgrades for its aging sewer pipes and dilapidated wastewater treatment plant. (*Courtesy of the San Francisco Public Utilities Commission*)

Sandbag Saturdays

One of the more proactive outreach activities was "Sandbag Saturdays." In San Francisco, the sewer system often floods during heavy rains. Particularly vulnerable are areas of the city that were built over old creek beds and other low-lying areas. Damage from these floods ranged from minor nuisances to more serious cases, such as basements of homes being knee-deep in raw sewage.

So, in the interest of helping residents to help themselves, the communications team proposed "Sandbag Saturdays." At the beginning of the city's rainy season, the utility sponsored a day of flood preparedness and free sandbag giveaways in four flood-prone locations throughout the city. The team believed it would help raise awareness of the sewer system's problems and the fact that it needed to be fixed. City residents praised the initiative as more than 1,000 sandbags were distributed to San Francisco households, and it received very positive coverage in the media.

Courting the Media

Until now, the utility had tried to avoid talking to the media. When it did, it was usually on the defensive, explaining a problem, such as the explosion at the Southeast Treatment Plant or a sewage overflow into the basement of someone's home or into San Francisco Bay.

Now, as part of their educational campaign, they courted the media, offering members of the press tours of the city's 150-year-old brick sewers and its treatment plants. Within a year, every local TV and radio outlet had taken them up on this offer and they broadcast stories about the conditions in the sewers. The San Francisco sewer operators became celebrities, featured on national TV shows like *Oprah* and *Dirty Jobs*.

These efforts paid off. In the beginning of 2007, the utility's general manager stood before San Francisco's Board of Supervisors and asked for a four-year rate increase for the sewer system, amounting to a double-digit increase for each of the four years. There was near-unanimous board approval, and only a handful of citizens spoke out against the rate proposal. By early 2008, when the utility's management presented its ambitious plan to retrofit the dilapidated sewer plant and aging pipes, there was very little criticism of the potential $4 billion price tag. This time, the media was on board: The city's newspapers published editorials supporting the retrofit.

The work of the utility's communications team provides a valuable lesson on the importance of fully educating and engaging the public. Many of the older sewer pipes and pumps are being replaced. The early results of pipe repairs are evident with the elimination of "Sandbag Saturdays." This outreach program is no longer needed as the sewers now flood less often during the heavy rains. In 2009, the utility moved forward with its sewer treatment plant retrofit.

As the plans and construction on the plant retrofit move forward—a project of this magnitude will take 10–12 years to complete—the utility continues to involve the community in the discussions about the "City under the City." It is planning to keep the city's residents engaged even after the retrofit is completed. In fact, one of the important design considerations of the retrofitted sewage treatment plant is that it also be an environmental educational venue and a destination spot for local residents.

This San Francisco case shows the importance of involving the community in a well-developed urban setting. In the next case, we are not concerned with repairs and upgrades; instead we are focused on providing the basics for public health. The following story illustrates how Brazil informed and involved its residents in its efforts to bring water and basic sanitation to its city slums.

EDUCATION AND ENGAGEMENT DELIVERS RESULTS FOR BRAZILIAN SLUM DWELLERS

Limited access to clean drinking water and basic sanitation has long meant illness and sometimes death for the residents of the *favelas* (slums) of Brazil. Fortunately, in many of Brazil's favelas, an unorthodox approach was successful in delivering clean water and removing sewage. An essential component of this process was the education and active involvement of the affected residents.

In Brazil, and in many developing countries, the large cities are often surrounded by slums, filled with families who have migrated from the rural countryside, often in search of work. Most of the residents of these fast-growing settlements live in dwellings that can be best described as shacks, which spring up around the

By involving Brazilian residents in the effort, the Brazilian government was able to deliver clean water and provide basic sanitation.

cities in a haphazard fashion. These slums often lack access to safe drinking water, proper sanitation, or both. In one of Brazil's slums, there was clean running water but no sewage system, so the human waste ended up in the ditches near the homes. Another settlement did not have clean running water, so the residents had to either drink contaminated water or buy very expensive water from water trucks. These conditions exacerbated health problems in the close quarters of the slums. Children are

"The toll on children is especially high. About 4,500 children (worldwide) die each day from unsafe water and lack of basic sanitation facilities. Countless others suffer from poor health, diminished productivity and missed opportunities for education."
—UNICEF—Water, Sanitation and Hygiene—
Children and water: global statistics (2006)

often the most directly affected victims of frequent bouts of severe diarrhea and, less frequently, outbreaks of cholera.

By the early 1990s, with increased migration to the *favelas*, Brazil's federal and state governments were struggling to find a solution to the problem. They were drawn to a system, called the *condominial*, developed in 1980 by an engineer named Jose Carlos Melo, that would allow governments to provide access to water and sanitation at a much lower cost than the conventional methods. But Melo's *condominial* approach was more than a new method of construction; it relied upon establishing a link between technology and the community.

The technological aspect of Melo's *condominial* approach is fairly straightforward. In a conventional sewage collection system or water distribution system, a network of pipes is connected to each home. The main pipes are buried deep underground and often down the middle of the street; these connect to smaller pipes, also buried deep underground, that carry water to each home or collect sewage from each home. By contrast, in Melo's condominial system, the conventional main pipes (or trunk lines) run only to each neighborhood or block, at which point a network of smaller water or sewer pipes is connected at much shallower depths under backyards, front yards, or sidewalks, to reach the individual homes. The diagram in Figure 4.2 shows the difference in the two systems.

The costs of the condominial network are much less expensive than the conventional approach because the conventional approach requires

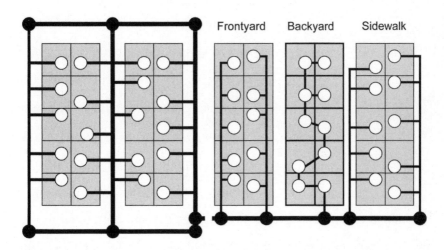

Figure 4.2 The conventional system (shown at left) has pipes buried deep underground, going down each street. The condominial system (at right) has only a connection at each neighborhood or block, with a shallow network of pipes branching out from it.

deeply buried pipelines to avoid accidental breakage, whereas the condominial approach allows for shallower pipelines and, therefore, less excavation. In addition, in the condominial approach, costs are reduced because the community is required to participate in the pipe placement, ongoing inspection, and maintenance of the neighborhood network of pipes.

Melo knew he needed the participation of the residents for the system to work. So, he gave his engineering approach a name that would resonate with local residents. He elected to call the concept *condominial* after

A condominial *is a sewer and water engineering concept named after a popular* telenovela.

> *The success of the condominial projects varied in direct proportion to the level of community participation.*

the name of a popular *telenovela* (TV soap opera). Coincidentally, the word *condominio*, or condominium, is used to describe a group of dwellings. The term "condominial concept" suggests a connection to an apartment or condominium building, but the definition is broadened to convey a group of dwellings that are horizontal to one another and comprise a neighborhood block. In Melo's approach, the water and sewer systems are developed to work in a block-by-block system.

Since 1990, there have been several *condominial* projects in Brazil. We describe three of them: one drinking-water project and two sanitation projects.

Brasilia: An Early Success Story

Brazil's first *condominials* grew up in the favelas around its capital city, Brasilia. Situated in Brazil's Federal District (the seat of the Brazilian government) in the Central-West region of the country, Brasilia has a total population of 2.5 million people. By the early 1990s, the population of Brasilia's favelas had grown to approximately 500,000 people. The residents of the Brasilia favelas were fortunate: their houses had running water and electricity, supplied by the government. Many houses also had toilets; these, however, were not connected to a sewer system, and toilet waste ended up in ditches near the dwellings, posing a serious health hazard to residents. By the mid-1990s, Brasilia's water and sewer utility knew it needed to address this public health problem. With limited money to create a conventional sewer system in the favelas, the utility turned to Melo's approach as a potential solution. The utility staff began by familiarizing themselves with the *condominial* approach. It had been implemented, albeit on a much smaller scale, in other areas of Brazil, so they could see how it operated, and it was agreed that it could work in Brasilia.

A community education campaign to engender the support of the favelas' half-million residents was essential. The utility began their efforts with intensive meetings with local politicians, including the governor of the Federal District, to explain the *condominial* system and get their support for the project. One of the first questions the utility was asked was: Why not install a conventional system—which we know works—instead of this new type of system? After providing detailed explanations of the system's benefits, the utility gained the support of Brasilia's elected leaders. Now they had to convince the favelas' residents.

Having the support of the local authorities helped deflect some of the residents' concerns, but residents of the favelas were skeptical. The *condominial* system was different from the conventional system and it was cheaper, so they assumed it was inferior. The utility embarked on a full-scale, grassroots outreach program that moved with the efficiency of a well-run political campaign. They held meetings with community leaders in each locale to explain the system and present the design options for each of the neighborhood blocks. The residents of each neighborhood elected a representative who would act as the liaison between the utility and themselves.

Electing the *condominial* leaders was just the first step in bringing the new sanitation system to the favelas. There was still a lot of work ahead before installing the pipes and getting the thousands of residents connected. Each *condominial* resident had to sign an agreement with the utility, confirming their commitment to participate in the system's upkeep. Each block's residents also had to make a collective decision about the design, whether they wanted the pipes installed under the backyard, front yard, or sidewalk.

Since many of the favela residents had little education and some were illiterate, information had to be conveyed in simple terms and there had to be several public forums to allow the residents to exchange ideas and ask questions.

Brasilia's effort to promote the development of condominials *involved* 5,000 public meetings, *attended by more than* 57,000 residents.

This intensive communication effort paid off: 100 percent of the favelas' 500,000 residents participated in the condominial sewer system, developing strong social networks in the process.

Bahia: A Case of Lessons to Be Learned

The people of the favelas of Bahia's capital, Salvador, live in dismal conditions perched on the hillside above the naturally beautiful Bay of All Saints. Like Brasilia, Salvador's favelas lacked a proper sewer system. But, unlike what happened in Brasilia, city officials in Salvador essentially ignored the unsanitary conditions until the sewage draining into the Bay below began to have a negative impact on beach conditions. The Bay's recreational

Brazil's effort to expand sanitation was less successful in the state of Bahia, but there are lessons to be learned.

beaches draw thousands of tourists to Bahia each year, and in the early 1990s, concern over the loss of tourist dollars moved the local authorities to take action. The state government then pulled together monies from several sources, including international development banks. The effort to bring sanitation to the favelas was then turned over to the Bahia Water and Sewage Company, a private company granted the exclusive concession to run the state's water supply and sewer systems.

From the outset, bringing improvement to Bahia's favelas was going to be more complicated than the project in Brasilia. The population of Bahia's favelas was close to 1.25 million, nearly three times the size of Brasilia's, and the ramshackle houses—which were crammed together and built precariously on the steep hillsides—promised to make the installation of any sewage system difficult.

As in Brasilia, gaining the support of the local residents was a key ingredient, but unfortunately, the water and sewer concessionaire did not place the same value on engaging and communicating with the slum dwellers. In stark contrast to the communication effort in Brasilia, the utility, the Bahia Water and Sewage Company, made limited attempts to educate the residents about how the *condominial* system worked and what their involvement in it would be. The lack of communication resulted in a slow and haphazard acceptance of the program.

Nonetheless, the *condominial* sewer program moved forward in fits and starts through several sections of Salvador, although it did not really pick up speed until 1997. After several years of effort, by 2004 the *condominial* sewer system was serving only 30 percent of the houses. What is most heartbreaking about this low acceptance rate is the lost opportunity to improve the health of the children in the slums.

Where the sewer system was put in place, there was significant decrease in severe diarrhea suffered by children under the age of four.

According to a study published in the British medical journal *Lancet* (2007), all the communities that installed a *condominial* sewer system experienced a 26 percent decrease in cases of severe diarrhea among children. In most impoverished areas, the rate of improvement was even more significant, with some areas experiencing a 43 percent decrease in illness.

Parauapebas: Using the *Condominial* Process to Deliver Safe Drinking Water

In 1996, a smaller but very ambitious project began to deliver drinking water to Parauapebas in northern Brazil using the *condominial* process. The water project was centered in Parauapebas, a town developed around the mining of iron ore. Parauapebas was first incorporated as a town in 1988, growing from 20,000 residents to close to 80,000 by the mid-1990s, to its current population of 152,000 (2009). As the town's population grew, its basic infrastructure did not keep pace. The water system serving the town's residents was hodge-podge at best. By the early 1990s, only a small portion of the town's residents had adequate water service. Most of the residents were using untreated water from the river or from contaminated wells. The lack of a water supply system and a limited sewer system resulted in widespread incidents of disease, especially gastrointestinal ailments.

By 1993, an agreement was reached between the major mining company in Parauapebas and the local authorities that, together, they would seek loans to finance the construction of a water treatment plant and water distribution (pipeline) system. The water treatment plant was constructed by 1996 but there was not enough money left to build a conventional water distribution system to transport the water from the treatment plant to homes. The most viable alternative was to construct a water distribution system using the *condominial* model, which would cost much less than the conventional water distribution system. But there was not enough money available to pay for the *condominial* system.

The local authorities were forced to ask the residents to help pay for the cost of materials and to assist in the construction of the *condominial* pipelines.

Originally, local authorities and the mining company had set out to educate residents on the proper use of the water system. Once the plans changed to require equity and "sweat equity," their community mobilization efforts were expanded to include three important steps.

First, there was intense dialogue with the community providing an exhaustive explanation of the workings of the *condominial* water network and why it was necessary to have residents participate in the construction. Residents were assured that the *condominial* approach would not cost more or be less effective than the conventional approach. This dialogue took place in numerous grassroots venues, including churches and political organizations.

Second, rules were established for splitting the construction work between the municipality and the community. The municipality agreed to be responsible for tasks that required technical expertise, such as determining the hydraulic capacity of the pipes. The residents would be responsible for digging the trenches, connecting the pipelines, and performing the ongoing maintenance.

Third, they began with a demonstration site in a small section of the city to build confidence in the project and serve as a model for the entire project.

The Parauapebas community mobilization efforts also included many of the components from Brasilia, such as holding meetings on the *condominial* level, the election of a *condominial* representative, and the requirement that all members of the *condominial* sign agreements pledging their support for the project.

This community mobilization effort concluded with 60,000 people in Parauapebas dividing themselves into 800 *condominial* units. The successful mobilization resulted in a significant increase in the number of families with access to clean drinking water from 18 percent to 80 percent.

> *The intense community mobilization in Parauapebas reaped stunning results. Access to clean water increased from 18 percent in 1998 to over 80 percent in 2004.*

Melo's Engineering and Community Engagement Approach Reap Benefits Half a World Away

The impact of Melo's innovation traveled well beyond Latin America. Half a world away in Asia, Melo's work was brought by one of his students and took hold in Karachi, Pakistan. In Karachi, there is a collection of 18 towns forming a settlement called Orangi; it is believed to be the largest informal settlement in Asia. One of the problems plaguing this settlement was the absence of a sewer system. Sewage flowed in open troughs in the lanes that wind through the settlement. In the early 1980s, a nongovernmental organization, the Orangi Pilot Project, came together with the goal to organize the settlement's residents to find a means for providing sanitation for its residents. With guidance from the Orangi Pilot Project, the residents of Orangi began organizing themselves to bring sanitation to their community. They were assisted in their efforts by Gehan Sinnaramby, a Sri Lankan who had worked with Melo on one of the first *condominial* projects in Brazil in the early 1980s. The Orangi Pilot Project effort was similar to that in Parauapebas, Brazil, where much of the funding and labor for the project was provided by the residents. In Orangi, the sewer project was self-funded, self-administered, and self-maintained, using the Orangi Pilot Project to coordinate activities. The sewer system had many of the same attributes as the *condominial* system and many of the same organizational traits, such as electing a "lane manager" (similar to a *condominial* representative) to represent about 15 households. By the early 1990s, the project was able to provide sewer service to over 600,000 Orangi residents. In Orangi, the government did not take the lead in the effort and, in fact, only after the Orangi sewer network was in place did the government bring in the trunk lines to connect to the municipal system.

Though very different places, San Francisco, Brazil, and Pakistan each developed a plan appropriate for their respective communities. While their success stories are a reason to celebrate, they also provide a cruel juxtaposition to cities in other developing countries. For example, in New Delhi, India, slums called *bustees* have sprung up around the city. As with the favelas, the bustee residents have come in from more rural areas, often looking for work, and have set up illegal settlements. Families in the bustees often live in tarpaper huts, and because the bustees are illegal settlements, they receive no water or sanitation services from the local government. Residents of the bustees must walk up to one-half mile to get access to water. And there are no toilets or latrines anywhere in the bustees. As a result, the residents—and there are close to four million—must resort to urinating and defecating openly in public. In such conditions, the health and the social fabric of the community are constantly under assault.

We have illustrated what can and has been accomplished. These solutions can be applied to slum settlements throughout the world. As the Brazilian case studies show, the technology can be provided for a fraction of the cost of conventional systems. These cases reinforce the premise that consumers, once informed, become involved and are willing partners in protecting their public health and environment.

Orange County, California, and Singapore:
Breaking the Psychological Barrier, Changing
Mindsets and Perception

In Chapter 2, we discussed the success of Orange County's recycled water program. We believe that it is also important to understand that its successful implementation was closely linked to its communication outreach program. It was a herculean effort to get the consumer, the county ratepayers, to be first in the nation to agree to get their drinking water from

> *Orange County waged a successful battle against the mantra of "toilet to tap."*

treated, recycled sewer water. Orange County's effective dialogue with the public and the media is in marked contrast to its neighboring counties, Los Angeles and San Diego, where many of the elected officials actually contributed to the hysteria over "toilet to tap."

In other California counties, attempts to implement a recycled water program were met with jarring opposition. In 2007, when Orange County was in the final stage of opening its recycled water plants, the City of San Diego began to seriously consider developing its own system for using recycled water. Though San Diego's water need was similar to that of Orange County, the response from the city's opinion leaders was mixed at best.

The San Diego newspapers railed against the proposal. An editorial in the largest newspaper in town (the *San Diego Union-Tribune*) headlined, "Yuck! San Diego should flush 'toilet to tap' plan." The editorial continued: "Your golden retriever may drink out of the toilet with no ill effects. But that doesn't mean humans should do the same. . . . You know the zealots behind the 'toilet to tap' initiative are trying to put something over on you when they change the name to 'reservoir augmentation'—a euphemism intended to obscure the nasty fact that the project would take heavily contaminated sewage water, purify it, and send it through your tap for human consumption."

San Diego's Mayor Sanders also attacked the plan: "I'll oppose any effort to bring about 'toilet to tap' . . . There is neither the money nor the public will to support such a program." As described by Maureen Stapleton, the head of the San Diego County Water Authority, "It turned into a big fat political football," referring to the battle between the mayor and the city attorney.

In Orange County, the discussions about recycled water were more reasoned and rational because, as the following case describes, Orange County avoided the political drama by thoroughly educating its consumers.

Orange County prepared for the worst. The water utility hired General Norman Schwarzkopf's former public affairs officer to head up its communications effort.

In 1998, as the elected board of the water utility decided to move forward with the recycled water program, they knew they would have some serious convincing to do. As they began the planning for this large infrastructure project that would take close to ten years to complete, they knew that it was not too soon to begin educating the people that would pay for the project and drink the water. The utility began its communication effort by hiring Ron Wildermuth, a retired Navy captain who had been in charge of public affairs for General Norman Schwarzkopf before, during, and after the Gulf War. A top-notch communications expert, Wildermuth was experienced in dealing with elected officials, the media, and the public.

With the support of the utility's staff and its board of directors, Wildermuth developed an ambitious communications plan that included meetings with elected officials and other opinion leaders at the federal, state, and local levels. Many of those leaders provided letters of endorsement and support. The utility also won the support of community and religious leaders.

They used all possible communication vehicles available: surveys, focus groups, and meetings—plenty of meetings. The program involved hundreds of face-to-face meetings and presentations. According to Mike Wehner, the utility's assistant general manager, "As more people learned about the project, they became more comfortable."

The utility also went out of its way to be candid about the treatment process. "We always tried to be up front with people. From the very beginning, we said we were treating sewer water, we didn't try to sugarcoat it," explained Eleanor Torres, the current communications director of the utility.

Over seven years, more than $4.5 million was invested in the outreach and communications program with its customers. The investment paid off. By the time the recycled water was being pumped into the county's underground reservoirs, surveys showed that 70 percent of those familiar with the project had a favorable opinion of it.

NEWATER WINS PUBLIC CONFIDENCE IN SINGAPORE

When the Public Utilities Board (PUB), Singapore's national water utility, introduced its own brand of recycled water, NEWater, in 2002, its success in achieving widespread public acceptance was based upon a very well-planned and executed communications campaign.

> *"The most important element in a successful water recycling program lies not in assembling the best technology or ensuring the quality of the water, but in winning public confidence and acceptance."*
> —*Mr. Yap Kheng Guan, a director with PUB,*
> *Singapore's national water agency*

The science and technology behind water reuse is only the first step; the real challenge is in overturning the stigma and the public's fear of reusing sewer water as a source of drinking water. PUB shifted the public's attention away from the source and focused instead on the treatment processes to show people that it was perfectly safe. They made a concerted effort to convey the technical information in simple layman's terms.

Unlike Orange County, PUB tried to avoid using words like sewage or wastewater because of the negative connotations that remind customers of the source and adds to their psychological fear.

From the start, PUB avoided the problems that had plagued other water recycling projects. Although it avoided using such "hot button" terms as "sewer" and "wastewater," it made every effort to be open in addressing all the public's health, safety, and quality concerns. And, early in the process, it focused on engaging the key stakeholders, especially the media and decision makers. Other cities that had promoted recycling wastewater like San Diego and Tampa, Florida, withdrew their proposals in the face of strong public opposition. PUB felt that these cities had erred by not beginning

> *The goal in Singapore was to make the public understand that this water is technically not wastewater but "used" water, and it can be used and reused over and over again, similar to how water recycles itself in nature without compromising quality and safety.*

their communication efforts with getting key opinion leaders to understand and accept wastewater recycling. PUB sought and won the support of public leaders, including the prime minister.

Changing the Terminology

In order to break through the psychological barrier to public acceptance, PUB made deliberate attempts to rename terms that had negative connotations with more neutral terms. Wastewater and sewage were referred to as "used water" to avoid the association with dirty, germ-laden sewage. "Used water" also better reflects its true value and significance in the water cycle as water that can be used and reused, similar to how water recycles itself in nature, and therefore is technically not wastewater to be thrown away anymore. Sewage treatment plants were renamed "water reclamation plants" because they no longer treated sewage; they were now engaged in conservation, by reclaiming the used water for further reuse.

One of the eye-opening aspects of the PUB campaign was explaining to the public about "unplanned indirect potable use." This involves a process by which the treated used water is discharged into the rivers, and then captured downstream for community use as tap water. This has been practiced by cities in Europe for centuries and in North America for over a hundred years. The public is generally unaware of this, and most people continue to believe that their tap water is drawn from "pristine" natural sources, such as high mountains, lakes, and reservoirs. In fact, most of our raw water is sourced from polluted lakes, rivers, and streams, and is only made potable through undergoing increasingly complex physical, biological, and chemical treatment processes (as described on pages 16–17.)

Singapore's Prime Minister: NEWater Champion; Field Trips for the Media

Singapore's prime minister became more than a supporter of NEWater (the recycled water); he became one of its champions. He and other government officials were often seen at public events or on television drinking NEWater from its distinctive bottles.

To underscore that water recycling is not a new concept and that it was happening worldwide, PUB briefed the news media and arranged to send reporters on trips to cities in the United States in which water recycling programs were up and running, such as Orange County, California, and Scottsdale, Arizona. The resulting media coverage helped assure the residents of Singapore that the technology worked.

PUB's communications and media strategy proved to be successful in Singapore. The NEWater has been widely accepted, and PUB's NEWater Visitor Center, built to showcase Singapore's recycled water process, is a popular destination for locals and tourists alike.

IN THE CASES WE have just described, governments, private companies, or public water utilities did a great job of reaching out to their customers. And their efforts are a model for anyone in charge of a water or wastewater system.

But what about all of us, the consumers, what is our role? At the beginning of this chapter, we asked, "Whose water is it?" Yes, of course, it's our water. But we often act as though it is somebody else's fault if water is not there when we turn on the tap or we are upset when our sewers overflow and pollute our rivers.

We assume such matters are the job of the utility or the government to deal with and inform us. There goes the assumption that we are a sophisticated, well-informed bunch. We implore you to become knowledgeable. Find out who is in charge of your water. Who is making sure that we are not contaminating our drinking water or our environment? Remember, it is no longer out of sight or out of mind.

CHAPTER FIVE

VALUING AN EXTREMELY COMPLICATED RESOURCE LEADS TO WISE USE

In December 2008, the chairman of the global food conglomerate Nestlé said, "I am convinced that, under present conditions and the way water is being managed, we will run out of water long before we run out of fuel." Implicit in his comment is the realization that energy resources are managed better than water resources. Collectively, as we have come to regard oil as a limited resource, we have become more efficient in its use. Water, on the other hand, has not historically been viewed as a scarce resource. Yet, in order to manage a resource well, we must understand and quantify its value. This task is particularly difficult with water because it is freely available in nature, seemingly abundant

> *To manage a resource well, it is important to understand and quantify its value.*

(until there is a drought), and seems to belong to everyone. Water is also very hard to catch and store, and after it has been used, it drains "away."

The value of water has been pondered by scholars for millennia. Plato observed that "only what is rare is valuable, and water which is the best of all things . . . is also the cheapest." Two thousand years later, in considering the difference between the market price of commodities and their economic value, the eighteenth-century economist Adam Smith compared the value of diamonds and the value of water. In his book *The Wealth of Nations* (1776), Smith made the distinction between *value in use* and *value in exchange*. Water, which has great value in use, often has little value in exchange, whereas diamonds, which have little value in use, have enormous value in exchange. Both Plato and Smith pointed out that the market price of an item did not always represent its true value.

In many countries water use is free, its capture and distribution being subsidized by the government; in other countries, water is captured and appropriated by individuals and groups for their own gain. Hence, there is a fundamental need for establishing a generally accepted and truly holistic perspective on the ownership and value of water.

SEARCH FOR THE VALUE OF WATER

How can we value a resource that has so many different attributes and uses? There are many conflicts between people who live in the upper reaches of a watershed and those who live in the lower reaches. These *upstream* and *downstream* users argue over removing water from its source, using it for waste disposal, or for the harvesting of fish and other species. Traditionally, these disputes were resolved by negotiation among tribal elders or by the threat, or actual use, of force. Over time, with the development of powerful centralized governments, nations assumed the role of arbiter, deciding who got how much water, and when, within their own borders. However, international conflicts still required negotiation. Some countries have experimented with modern economic bargaining techniques, such as water trading or the sale of water rights. Such negotiations typically led to the assignment by the governments involved of some sort of property right over irrigation water and to the creation of markets where those rights can be bought, traded, or leased. Australia, Chile, and to a lesser extent, the arid Southwest of the United States are the leaders in the economic allocation of water. For domestic and industrial water supplies, most urban water system facilities are run by public authorities. Under both domestic and industrial water supply systems, there is a need to be able to measure the value of water to each different user.

WHAT IS THE ECONOMIC VALUE
OF WATER AND HOW IS IT MEASURED?

Meaning of Economic Value

Can the economic value of water be measured by its market price? If that were true, then only marketed quantities of water would have economic value. But in the many parts of the world where domestic water is not sold or marketed, it still has great economic value. Even where water is sold, either in bulk to industry or retail to consumers, its actual economic value typically exceeds the price charged for it.

> *Around the world, consumers pay more for water as its scarcity increases.*

In discussing the value of water, economist Michael Hanemann at the University of California, Berkeley, made three key points, which generally hold true for all commodities: First, *demand* indicates what a good is worth to people at what price, and *supply price* indicates what that good costs to produce and deliver; second, the *market price* reflects the interaction of demand and supply; and third, how people value a commodity reflects their *subjective preferences*. So, for instance, some consumers may be willing to pay a huge price for a small quantity of water if they are in a desert and this is the last cup of water available. Under different circumstances, however, a consumer may only be willing to pay a modest price for a bucket of water, and perhaps even less for a bathtub full. A *demand schedule* is this relationship between price and quantity, and it reflects a behavioral trait of consumers. Consumers in Kansas behave pretty much the same way as consumers in Kathmandu—the price they are willing to pay for water decreases as its availability increases.

The supply price is the cost of producing an additional quantity of water. The water supplier develops the cheapest water sources first, such as a nearby river, then the next cheapest, and so forth. This implies a rising supply-cost curve, or *cost of supply schedule*. Using the principles of modern classical economics, it is possible to compute the total amount a consumer is willing to pay for marketed water based on the demand schedule and the cost of supply schedule, and this should tell us the real value of water in any given situation.

Valuation of Nonmarketed Goods

When seeking the value of a water product, there is a distinction between water itself as a commodity for consumption (discussed in the previous section) and the services provided by water (discussed here). Water services are the measures of the uses to which water is put; for example, draining away polluted water and sewage, or as a medium for fishing, swimming, and recreation. In natural settings it is difficult, if not impossible, to charge a price for these activities (or services rendered). They are not without value, but there is no simple market in which they are sold and purchased. For some of the services, such as swimming, we can assess the value by comparing it with commercially run swimming pools. Similarly, water-based recreation can be compared with similar commercial activities. Even for wastewater disposal, there are alternative systems for which people pay, such as septic tanks, and the prices charged could be a good proxy for this service supplied by water.

Then there are the unpriced effects of upstream-downstream externalities. An *externality* is monetary damages inflicted on other users of water services by the action of one individual. So, if an upstream farmer or community diverts water from a river, then the downstream users and communities suffer economic damages by the lack of access to water for their own uses. The existence of externalities does not always imply upstream-downstream impacts. Externalities could stem from one farmer over-pumping groundwater, which will necessitate all other groundwater users in the vicinity to have to spend more money in pumping from greater depths. The fact that these external effects are not incorporated in the prices paid by the offending parties for their water use challenges the classical Adam Smith view that the market can provide the correct value for water and water services.

The U.S. Army Corps of Engineers is charged by Congress to ensure that the benefits from publicly provided, multipurpose water projects exceed the costs of providing them and to assess these before the project is actually built. For the Corps of Engineers, evaluating the costs is quite straightforward; they just add up what they know about the costs of building and managing the project over its lifetime. For multipurpose projects, the benefits from some of the purposes, such as hydropower generation, have clearly defined market prices and, hence, value; for instance, electric energy has a well established market price. The difficulty is that other purposes, such as water-based recreation in remote areas, do not.

There are, however, ways to find out what a consumer is willing to pay, or has actually paid, for a nonmarketed item, using comparable marketed activities, as mentioned above. One approach was developed and used extensively by the U.S. Army Corps of Engineers in the 1950s to estimate the benefits of publicly provided water projects. For example, to assess how much visitors would be willing to pay to stay at a water-based recreational facility, the Corps developed the *travel-cost method* for evaluating the benefits of such recreation. Recreation-seeking individuals, using Corps of Engineers–managed reservoirs, were given questionnaires and asked to indicate how much money they had spent traveling to and using the water-recreation facility. These costs, summarized over all potential users, were then used to estimate their willingness to pay for the unpriced water services at the recreational facility, and, hence, the value of the water services.

Other methods were developed to assess consumer preferences for unpriced commodities. *Contingent valuation* is a survey-based economic technique in which people are asked how much money they would be willing to pay to maintain an environmental feature, such as water quality in a river, or wetlands to protect wildlife. *Hedonic pricing* is a method of estimating the value of nonmarketed water services in relation to their economic contribution to increasing the value of commodities with well-defined markets, such as real estate. Importantly, the nonmarket valuation methods apply to both the positive and negative outcomes associated with a water project. That is, the assessed value of a water service, such as wastewater emissions, will cause a reduction in the value of marketed goods (again, property values will be depressed along the shore of a polluted body of water). In the United States, water agencies routinely use these methods to assess the environmental damages due to water projects.

HOW IS WATER DIFFERENT
FROM OTHER COMMODITIES?

Water as a Private Good and Water as a Public Good

Economists like to draw a distinction between conventional *private goods* (goods owned by individuals or groups) and *public goods* (goods held in common). The main distinctions between the two are based on whether these commodities exhibit *non-rival consumption* and *non-*

excludability. For a commodity to be a public good, it is not possible to exclude other consumers from consuming the good, and their consumption of the good must not interfere with any other participant's consumption (*non-rival consumption*). In the area of water management, one such public good is flood protection—once an embankment is built, you cannot exclude persons behind the embankment from enjoying the benefits of flood protection, nor does their enjoyment of flood protection reduce in any way the availability of flood protection for others. On the other hand, well water on private property is considered a private good: The owner of the well can forbid others to draw water from it, and if they do draw water, it reduces the amount available for the owner, thus interfering with the owner's consumption (*rival consumption*). Unfortunately, particularly with water, there is an "in-between" category: the *common-property resource.* Common-property resources combine the public-good attribute of nonexcludability and the private-good attribute of rival consumption. Water in a lake or a river is a typical example of a common-property resource. It is difficult, if not impossible, to exclude others from using the lake for, say, fishing, but this reduces the number of fish available to other users. So, access to the use of the water is nonexcludable, but in this case nonexcludable use also allows rival consumption.

One of the major problems with managing water as a commodity is that it moves from public to private to common-property status almost at will. For example, access to a municipal water supply is not a public good, because people outside a given pipe network cannot access the water from that network; in other words, they can be excluded from publicly supplied water. On the other hand, it is often not possible, for political reasons, to exclude people once they are within the pipe network system. At the same time, consumption by any one consumer on the network reduces the total amount available for other users. The combination of nonexcludability (public) and rival consumption (private) makes it act as a common-property resource.

> *Even if water is considered a public good, there are significant costs for delivering water and operating water and wastewater systems to protect public health and the environment.*

Much of the misleading rhetoric about provision of water-supply services stems from the widely held misconception that *publicly supplied water* is a public good. Many environmental activists, for example, assume that an undeveloped water source is a *pure public good*, because it meets the criterion of nonexcludability (but not rival consumption). In a dangerous leap of logic, social activists insist that urban water supply should therefore be provided free of charge by the government. However, even if it were a public good, somebody still has to pay to convey the water from the source to the user. From the point of view of practical resource management, providing water to the public for free, or for tariffs that do not cover the full cost of the system, leads to underfunded water systems with no resources available for improvement and maintenance and inefficient operation. This has led to many cases in which the government has turned to the private sector to help improve the efficiency of operations and the quality of the water supply. There are a wide range of such public-private partnerships in the water and sanitation sectors, ranging from the private participation being restricted to collecting revenues, all the way up to owning the whole system outright.

Social activists have objected to the "commodification of water," and applied pressure on the international sources of funding of the large multinational water companies. Investment banks, such as the World Bank and the Asian Development Bank, that finance many of the water systems in the developing world along with bilateral aid agencies, such as the U.S. Agency for International Development (the branch of the State Department that deals with overseas development assistance), have been majorly criticized for turning water into a private good and encouraging multinational corporations to exploit water services for a profit at the expense of the local citizens who are usually extremely poor.

These complaints are often also joined with the move to proclaim "water as a human right," and some countries, notably South Africa, have already enshrined it in their laws. The conjunction of ceding property rights and human rights has unfortunate consequences for many poor people, because most of the badly managed and poorly funded municipal water supplies around the globe are those run by public authorities in countries that have no record of reliable public sector performance. In these countries, government ownership often comes with sloppy financial and technical management, overstaffing, significant losses of water from the system, and poor water quality. Unaccounted-for water ranging as high as 40 to 70 percent of the water processed in the system was noted in places like New Delhi, Jakarta, and Manila before its utility was privatized to improve its performance. This is not to deny that there have been serious problems with privatization of municipal water systems, such as Buenos Aires, Argentina; Cochabamba, Bolivia; and Atlanta, Georgia, in the United States. In each of these cases, the local governments sought private help to improve the management of their systems and took back responsibility for them when they felt that the privatization was not working as they envisaged. In each case they now have to figure out a way to deal with the problems that had led them to seek privatization in the first place, which are largely unresolved.

The poor are often subjected to poorly maintained and underfunded water and wastewater systems. They also pay many times more for water than more affluent customers.

It is important to note that whether water is supplied publicly or by some public-private partnership, two major issues have to be resolved: how is the present financial security to be maintained, and how is the utility to be regulated as a monopoly? Both of these issues are fraught with potential for disaster. How is the utility to be financed in an efficient and equitable manner? This is difficult because in many situations the revenue base for the utility is weak—because the customers are themselves poor. Top quality systems that achieve high levels of technical performance are very capital- and management-intensive. To keep a utility solvent under such conditions will require some form of subsidies; some will have to come from general governmental taxes, and some from cross-subsidies from the richer customers to the poorer ones. All water supply systems are natural monopolies and hence they need careful monitoring on both the financial side and the safety and public health side. This is unequivocally a government function; it cannot be farmed out to the private sector. In many of the cases in which privatization efforts have failed, the lack of rigorous oversight has been a major component of the failures.

The Mobility of Water

One distinguishing feature of water is its mobility; it is often called a "fugitive resource." This term denotes a previously unclaimed natural resource that wanders from place to place (e.g., oil and gas in the ground, and water in the air, in the ground, and along its surface). Water clearly moves around: it flows downhill, it evaporates, it seeps into the ground; and typically, it is rarely fully utilized by its original user. The fact that we have "left over" water following its primary use has important economic, legal, and social implications. Keeping track of water is complicated and often expensive. In addition, the fugitive nature of water makes it the ideal source of transboundary and up- stream-downstream conflicts. If nobody seems to own it, or there are no sanctions against taking it, then the upstream users will take as much as they please, and pollute it at their leisure.

The Variability of Water

In addition to its mobility, water is also highly variable in terms of space, time, and quality. There is huge spatial variability among re- gions and countries and there is the pervasive temporal and seasonal variability. For example, rainfall is highly variable within space and time. Twenty-four-hour television stations devoted to reporting the weather give testimony to the variability of the weather: Consider the many manifestations of extremes of water, in the form of floods, snowstorms, and droughts. The major challenge facing water agen- cies around the world is how to plan for steady and increasing de- mands for water in the face of the spatial and temporal mismatching of water supply and demand. Not only is the supply variable, but the demand is also often quite variable depending upon the season, the day of the week, the time of the day, and, of course, the price charged for water.

The Cost of Water

Water is a bulk commodity that is very heavy relative to its unit price. The transportation infrastructure for water—water mains, reservoirs, dams, sewer pipes, and the like—is therefore also large and bulky. It is expensive to transport and store water, to treat it up to potable standards, and then to dispose of the wastewater. As we noted in Chapter 2, in California 19 percent of the state's entire electricity use in the state is taken up by these activities. As a result, the water utility sector has an investment-to-revenue ratio that is 70 percent higher than that of the electricity and telecommunications sectors. In other words, the ratio of the amount that has to be invested to provide one dollar of revenue for water is 70 percent higher than the amount invested for one dollar's worth of electricity.

Moreover, as a result of its large and bulky investments there are economies-of-scale for water projects that make the tradeoff between short-term and long-term investment decisions more complex, typically leading to 50- to 100-year horizon investments. In addition, the costs-per-gallon-stored when there is existing storage capacity (in economic parlance, the "short-run marginal costs") are much lower than when it is necessary to build extra storage capacity (the "long-term marginal costs"). This, plus the high fixed costs of the capital expenditures make water suppliers "natural monopolies," which arise where the largest supplier in an industry, often the first supplier in a market, has an overwhelming cost advantage over other actual or potential competitors. Hence, there is a need for strong governmental regulation of both private and public water utilities. These arguments are the basis for support of strongly regulated public or private water services.

The Price of Water

In most developing countries, such as China and India, water is owned by the state and is made available to consumers at a very low cost or even for free. As a result, it is important to note that consumer prices around the world reflect only physical supply costs and not the *scarcity-based value of water.* Unlike oil and coal or other minerals, for water there is typically no royalty payment by users to the government for extracting the resource. Another problem with water pricing is the practice of the water utilities and governments of setting prices to cover historic costs and not future replacement costs, which will always be higher per unit than the historic ones. In the nineteenth century, economist David Ricardo outlined a theory based on "increasing costs," due to the lowest cost activities being used first and the more expensive ones having to be used as the demand grows or as projects wear out. This effect has

> *Water pricing often does not take into account the future costs of maintaining the system; yet, we depend on the system to deliver water far into the future.*

led to economics being labeled the "dismal science." Accounting for future costs based on historical costs will always lead to an underestimate of the cost; this leaves the water supplier perpetually scrambling for rate increases. If a forward accounting stance were taken into consideration by proper asset management, these future cost increases would be built into existing tariffs, leading to modest and slow price increases.

Water Is Many Different Things

Water is not a homogeneous commodity. As we pointed out above, water has many different attributes including spatial location, timing, quality, and variability. Hence, every liter of water is different if any one of its attributes is different. This implies serious estimation problems when examining valuation of water supplies. For example, on a river, the water available to downstream users is often contaminated with domestic, industrial, and agricultural wastes from upstream. This means that the same amount of water is less valuable to the downstream users, since they will have to pay to clean it up before using it. In another example, water in the mountains of Northern California is much less valuable to the local populations than the same amount used for irrigation and urban uses in Southern California. Even reclaimed water in Singapore is more valuable to industry than to domestic users because industry does not have to spend money to upgrade its purity, as it does with the traditionally supplied municipal water.

The Essentialness of Water

Finally, water is essential for life: human, animal, or plant. In addition to being a foundation of our existence, it is also an essential ingredient for almost all of our actions. However, beyond a minimum level of about 2.5 to 10 gallons (10–50 liters) per person per day, there is little about the essentialness of water that should influence our decisions one way or the other. South Africa is one country that has legally enshrined the right to minimum amounts of water as a human right, but even supplying this amount to all the citizens is turning out to be a difficult task. This is because of the high cost of building the infrastructure in many poor cities, which have never had adequate infrastructure in the past. Providing free water to the poor in large wealthy cities, like Johannesburg, is relatively easy since the possibility exists of cross-subsidies from the rich to the poor. Hence, the only question, really, is what is the monetary value of an improved water supply? Dale Whittington (1995) showed that in the developing world the willingness to pay for such improved supply was in the range of 3 to 5 percent of household income.

How to Set the Price?

One of the perennial solutions to resource scarcity offered by economists is marginal-cost pricing. This idea, based on classical economics, holds that in a perfectly functioning market (devoid of public goods and externalities) the *demand schedule* (which represents decreasing willing-ness to pay for larger quantities) and the *supply schedule* (which represents increasing marginal costs) intersect at the socially optimal price at which the market will clear (supply matches demand), leading to the optimum allocation of the resource—or marginal-cost pricing. But in water plan-ning and management, the concept of marginal-cost pricing has caused a lot of mischief. For example, marginal-cost pricing could lead to prices that would make water unaffordable for the poor, who then could not purchase adequate amounts of water for cooking, health, hygiene, and sanitation. Over the past decade there has been much unhappiness on the part of non-governmental organizations (NGOs) about the proper pricing of water. This has often led them to major confrontations with the multilateral aid institutions, such as the World Bank and the Asian Development Bank. The large international water companies, which provide water services in many countries and make money doing it by charging tariffs higher than had been previously charged by govern-ment water utilities that were providing much lower quality of service, are also under great social and political pressure to keep the prices low. This makes it very difficult for them to make enough revenue to cover their costs. The conflicts over water pricing by privatized water utilities have been very contentious, violent, and in the case of Cochabamba led to at least one death during confrontations between the protesting water consumers and the police.

At the same time, even in the developed countries, notably in the United States, water pricing has remained a sore political issue. In a democracy, any politician who plans to be reelected is unwilling to raise prices for water and other similar services, though the infrastructure may be on the verge of collapse and the management of entire systems is compromised by a lack of funds. Indeed, in one celebrated case, the entire elected city government of Phoenix, Arizona, was recalled on a citizen's initiative because they had raised the price of water supplied by the publicly owned water company. Shortly afterward, a "water reform" slate of city officials was elected. But when they examined the financial records of the utility, they agreed to the price rise that had been set by the previous city council. By then, the water consumers were unable to sustain their anger with the system and grudgingly accepted the new tariffs. However, in most other parts of the United States, water tariffs are so low that the demand appears to be inelastic—in other words, raising the price of water does not seem to decrease demand. Unfortunately, this is not the case. Rather, the price increases that are made fail to keep up with inflation, so that water becomes proportionately less expensive over time. Many utilities rely on other sources of financing, such as property taxes, general obligation bonds, and other subsidies to run the system and bolster the direct revenue from water sales. The lack of price responsiveness to water pricing has become part of the industry folklore, and water utility managers do not push the political system hard enough to get meaningful price increases.

In addition to revenue generation, pricing has also been proposed as a mechanism for conserving water, above and beyond the needs of revenue for full-cost recovery of a water supply. Using pricing for conservation is implicit in marginal-cost pricing; at the correct price the socially optimal allocation of water will be achieved. However, in many situations where the true marginal costs are not known, it is possible to raise prices just to reduce the demand to what seems practically feasible. This is called rationing-by-pricing. This only works when there are technical methods for conserving water that only become economically feasible as the price of water rises. In the following case, the pursuit of full-cost pricing led to huge reductions in the demand for water. The actual physical reductions in demand were caused by the adoption of cost-effective technical conservation methods.

RATIONING BY PRICE: THE CASE OF
THE BOSTON METROPOLITAN WATER SUPPLY

Sometimes, however, price does matter; at least it did in the Boston Metropolitan area. From 1986 to 2009, the price of water increased tenfold while the demand for water dropped by 35 percent.

The demand for water supplied by the Massachusetts Water Resources Authority (MWRA) has changed radically since 1986. Currently, there are over 2.1 million residents in the MWRA service area. The demand rose rapidly from 1959 until 1973, when for the first time

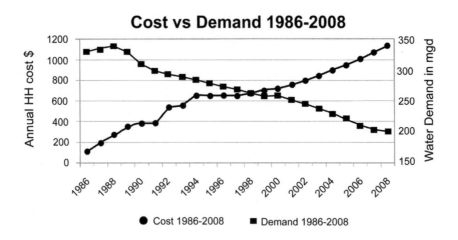

Figure 5.1   Chart from the Massachusetts Water Resources Authority showing the decline in water use with the rise in water and sewer rates. The MWRA also put in place a number of conservation programs, including mandatory low-flow toilets.

it exceeded the safe yield of the system, estimated at 300 million gallons per day (mgd). Because demand was expected to keep growing, it led to searching for new sources of supply. Various plans, based on transboundary diversions from the Connecticut River (which flows between Vermont and New Hampshire, through Massachusetts, and finally into Connecticut where it enters the Atlantic Ocean), were drawn up by the MWRA and its preceding agencies, with an estimated cost of up to $500 million. In response to the general concerns voiced by environmentalists about diverting rivers out of their basins, Michael Dukakis, the then governor of Massachusetts, signed the Inter-basin Transfer Act, which effectively placed a ban on transboundary diversions, leaving Boston to continue its upward drift of water demand. By 1986, they were exceeding the safe yield of the storage system by about 25 mgd. As a result, demand management programs began just before the MWRA was created, and were reinforced and accelerated once the MWRA decided not to push for the Connecticut River diversion. They were put in place to reduce demand, so that no new supplies would be needed.

The MWRA had been established in 1984 to address the major water planning problems in the Boston Metropolitan area and to take over the existing infrastructure and management and plan for future developments. Its first task was to address the deplorable state of Boston Harbor, which was the receptacle for the entire region's lightly-treated wastewater, including storm-water overflows that flushed large amounts of untreated waste into the harbor.

The saga of the clean-up of Boston Harbor is too long a story to cover in detail here, but the title of a 2004 book by Eric Jay Dolin says it all: *Political Waters: The Long, Dirty, Contentious, Incredibly Expensive, but Eventually Triumphant History of Boston Harbor—A Unique Environmental Success Story*. One can read it and weep, but it *is* eventually a success story. It is a tale about a malfunctioning state agency, the Metropolitan District Commission (MDC), gaps between federal and state enforcement of environmental regulations, an activist judge, the political temerity of elected representatives, and many decades of neglect of Boston Harbor by everyone involved. The MDC had been established in 1919 in a merger between the Metropolitan Water and Sewer Board and the Metropolitan Parks Commission. This odd marriage of parks management and wastewater management lasted 65 years, until the crisis of Boston Harbor led to the water and sewage units being split off to form the MWRA. Because of the MDC's mismatched functions and the fact that water and sewer customer revenues went to the Commonwealth of Massachusetts and not into the coffers of that branch of the MDC, the water and sewer division had to rely on allocations from the legislature to cover its expenses. As a result, the MDC was always out of money. There was no money to build new treatment plants and sometimes not enough to maintain the existing ones. The budget shortfalls also led to major staffing difficulties and maintenance problems, resulting in a nonfunctioning wastewater collection and treatment system that was supposedly protecting Boston Harbor. The problem was well understood by the technical staff of the agency, but the legislature consistently cut the MDC's budget for water and sewer management. By 1986, Boston Harbor was one of the most polluted harbors in the country.

THE FEDERAL CLEAN WATER ACT of 1972 was a wake-up call to the nation, detailing the appalling state of U.S. rivers and estuaries. Essentially, the federal government had allowed unrestricted use of its waterways as receptacles for all types of pollution. Because water has always been a states' rights issue, it was left to the individual states to regulate the pollution of their water bodies. Not surprisingly, the states had little motivation to enforce strict regulations that would cost local industries and municipalities money, particularly at a time when they were in competition with each other to entice industries to their state. This led to a "race to the bottom"; in effect, states were competing to lower, rather than raise, their water quality standards. Some earlier federal regulations, from the 1948 and 1953 clean water acts, eventually provided some minimum levels of regulations on all states, but essentially left the individual states to set their levels at least as high as the federal minimum levels. A major requirement of the 1972 act was that at least secondary treatment (removing about 80 percent of the biologically active material) be used for all publicly owned wastewater treatment plants. It would take 14 years before Boston came into compliance with the new regulations.

In 1982, in response to the not-so-benign neglect of Boston Harbor, Wollaston, one of the harbor communities, sued the MDC under the 1972 Clean Water Act to get the agency to clean up the fecal matter from untreated wastes that was fouling its beaches. This provided an opening for the state district court to enforce the federal law and for the arrival of Judge Paul Garrity on the scene. Garrity, a superior court judge appointed by Michael Dukakis in 1976, had a reputation for being an activist judge. In 1979, after the Boston Housing Authority, the city of Boston's largest landlord and the largest public housing authority in New England, was sued by tenants over unsafe and unsanitary conditions, Garrity had it placed into receivership and he supervised the day-to-day operations until the Authority was able to demonstrate that it was capable of complying with Massachusetts' sanitation and housing laws. Now, for the harbor lawsuit, Garrity appointed a Harvard Law School professor, Charles M. Haar, to report to the court that the judicial orders were actually followed. This was the turning point of the whole saga. From this point on, the major problem was that upgrading the existing wastewater treatment plant to meet the EPA's effluent standards would cost $3 billion. The unique funding mechanisms that were used at the time by the MDC meant that the legislature would have to appropriate tax money for this purpose—not a happy thought for the politicians. The MDC, looking for ways to avoid these huge costs, decided to ask the EPA for a waiver. Subsequent amendments to the 1972 Federal Clean Water Act provided that waivers could be granted to coastal cities, exempting them from meeting the EPA standards in cases in which they could use an ocean outfall to carry primary treated sewage (via a long pipe) out into the ocean. Several unfruitful years were spent by the MDC in negotiations with EPA to obtain the waiver, which was finally denied in 1985. The MWRA, which had been set up the previous year, assumed the debt of the MDC and set out to build a system in compliance with the law.

WHAT DID ALL THIS have to do with supplying water to the Boston area? A lot, since in Boston and all the contributing towns, consumers were charged for both the water supplied and the cost of treating the sewage they produced. What happened, and Figure 5.1 on page 144 shows this clearly, is that the $3.8 billion needed to construct a new treatment plant, the Deer Island Wastewater Treatment Plant, had to be raised by increasing the price of domestic and industrial wastewater treatment. Very rarely in the history of public utility management has there been such a steep increase in water/wastewater tariffs. After such a rapid price rise, one would expect to see a radical reduction in water consumption, and this is what happened in Boston. For a 20-year period, from 1969 to 1988, the customers of MDC and then MWRA routinely drew more water than the "safe yield," the annual amount of water that can be taken from storage 99 percent of the time without lowering the storage capacity.

It is instructive to look at how the average prices for a household using 90,000 gallons of water per year changed over time from before, during, and after the system was completed. Before the new treatment plant's construction in 1986, the combined average household bill for water and sewer was $113 per year. By 1991, it had climbed to $382; by 2000 it had reached $718; and in 2008 it was $1,132—a tenfold increase in 22 years! Note, however, that the total quantity of water demanded had dropped over 35 percent for the same period, down from more than 320 mgd to

Increasing costs led to rapid acceptance of conservation measures.

200 mgd. It certainly looks like the price of water supplied is actually quite elastic.

If water and sewer services in the greater Boston area were under-priced before 1986, how does the current price compare to the prices in other U.S. cities? The costs vary, but the average puts Boston in 2009 near the top of the list of expensive cities.

Price increases get people's attention. An effective agency marshals that attention, and focuses it on what consumers can do to control their costs, rather than simply leaving them angry. Moreover, price doesn't reduce demand, concrete specific actions on the part of consumers do. A good demand management program couples attention-getting activities with water-saving ones. The MWRA, on its website, attributes the demand reduction to its own demand management and water conservation programs, launched in 1986. The steepest decreases in water use were in the very beginning, as prices were just beginning to rise. And conservation expenditures were scaled back as demand dropped below the safe yield target. By 1989, withdrawals had been brought below the safe yield, and they have continued to decline ever since. The MWRA claims that the reduction in average water use was achieved through a vigorous program of leak detection and pipe repair. The program also retrofitted 370,000 homes with low-flow plumbing devices. A Water Management Program for area businesses, municipal buildings, and nonprofit organizations was established. Extensive public information and school education programs were developed. A change was made in the state plumbing code requiring the installation of 1.6 gallon per flush toilets in new construction and when replacing old toilets. Improvements were made in water meters that helped the participating MWRA communities to track and analyze their total water use. The MWRA also promoted new water-efficient technology that led to reductions in residential use. Finally, they initiated water pipeline replacement and rehabilitation projects throughout the MWRA and community systems.

Of course, it is hard to justify such programs when water is selling for $113 per household per year. However, when the prices rise sharply as in the Boston case, all these activities become cost-effective—water is only worth saving when it is a valuable commodity. The MWRA's efforts clearly fitted well into the governing economic theories; as price goes up, demand goes down provided consumers have some options for reducing their consumption of water. The MWRA leadership provided these options through their conservation programs and initiatives, and their fortitude in maintaining the increased price levels ultimately served the purpose of making the area's water supply sustainable well into the future. And it was the thoughtful leadership of certain individuals that helped propel the MWRA to do the right thing. The first was Governor Michael Dukakis, who in 1984 pushed the legislature to modernize the entire water and wastewater management of Eastern Massachusetts by creating the MWRA, and the second was the activist Judge Paul Garrity, who tenaciously monitored the MDC to keep it from avoiding its responsibility to meet federal laws and regulations.

Ultimately it is the consumers, who can understand price increases to provide the funds needed to make system improvements—they are less likely to be supportive if they cannot make that connection. The MWRA is happy to see demand reduction resulting from the higher prices needed to finance investments in clean, safe water, but it would not dream of raising prices solely to curb demand. Indeed, the MWRA is now suffering from its success. It now has a plentiful excess supply of water in its reservoir system and is seeking ways in which to market this water to ensure the financial sustainability of the institution.

WHAT CAN WE LEARN ABOUT THE ROLE OF ECONOMIC THINKING IN WATER MANAGEMENT?

The actual act of valuing water in a realistic manner leads to large changes in human behavior. A popular environmental slogan is "Reduce, Reuse, Recycle," which interestingly offers us a framework within which we can start to see how the component parts of this book—and the water cycle—come together. The message from this chapter should be clear from the MWRA case—demand reduction is a viable option for reducing the stress on our water resources. By being forced to raise the water price to cover the full cost of building and maintaining the water and wastewater system, consumers jumped at the chance to participate in water conservation programs because that saved them money.

However, this chapter also stresses the well-known fact that many people around the world may not be able to afford full-cost–priced water, and that subsidies may be necessary to provide them access to the protected water sources. In these cases one should not succumb to abandoning full-cost pricing, but rather temper the social exclusion by cross-subsidies from the wealthy to the poor. The second part of the slogan, "Reuse," was well covered in Chapter 2 by the cases of Orange County and Singapore's transformation to NEWater, which can be used directly as potable water. "Recycle" is taken up in other chapters, for example, in Chapter 6 in the section on the Geysers, where treated secondary sewage effluent is recycled as the working fluid in the geothermal fields. Hence, despite some misgivings about the usefulness of economic theory applied to a complex resource such as water, this chapter concludes that classical economic concerns, with particular attention to pricing, is one important part of the framework within which we should manage water.

CHAPTER SIX

WASTE NOT, WANT NOT

I t should be clear by now that wastewater disposal is a key component of all urban water systems. In order to properly dispose of wastewater (the used water), the operators of the sewer systems must focus on several issues, including protecting public health and the environment, energy use, and odors. However, as we saw in the San Francisco case, disposal systems are often considered the stepchildren of water utilities; most people don't want to talk about sewage. In this chapter, we explore the benefits that can be reaped from sewage and sewage treatment plants when forward-looking people are in charge. In each of this chapter's cases, the utilities adopted programs that helped to bring us closer to our goal of protecting our public health, environment, and water supply for our future.

FATS, OILS, AND GREASE CAUSE
HEART ATTACKS . . . IN THE SEWERS

Every year, millions of gallons of grease find their way into the sewers of most major cities. We all know what happens when fat-laden diets clog our arteries. Similarly, when large volumes of fats and grease from restaurants and fast-food shops clog city sewers, it is like a heart attack in our sewers. These sewers, blocked with grease, often overflow, and the result is a costly mess with environmental and public health consequences. In a recent report, the United States EPA stated that FOG (fats, oil, and grease) is the most common cause of sewer blockage (47 percent).

We described in Chapter 4 how the San Francisco water utility dealt with educating its customers. Here we look at the way it deals with FOG. The city's water utility has come up with an innovative solution that has turned FOG from an environmental hazard into an economic asset: clean biodiesel. The cleaner home-grown biodiesel will replace polluting foreign-sourced diesel fuel in all city-owned vehicles.

Restaurants + Sewers = Blocked Sewers

Throughout the developed world, cities spend millions of dollars trying to manage the FOG problem. Everywhere, the combination of sewers and food service establishments add up to problematic sewer blockages. Ask anyone involved in operating a water utility how they deal with FOG, and they will know that you aren't asking about the weather; you are talking about fats, oils, grease, and sewer overflows.

How does this grease end up in the sewers? The same way everything else does—through kitchen sinks and building drains. Without a concerted effort by food service establishments, FOG—tens of thousands of gallons of it each year—ends up in our sewers.

Figure 6.1   Striking image of an eight-inch sewer pipe clogged with fats, oils, and grease (FOG). The FOG, which builds up over months after being dumped into the sewer, can become as hard as concrete. The hardened FOG has to be either jackhammered out; or, in the case of a smaller pipe such as this one, the section of pipe is cut out and replaced. (*Courtesy of the San Francisco Public Utilities Commission*)

Most food establishments do make an effort to keep grease out of the sewers, but many are unsuccessful. It is not an easy task: it requires having a comprehensive system that includes training kitchen staff on proper disposal; using the proper grease collection containers and equipment; and, finally, paying to have the grease removed by a hauler approved by the local water utility or public health officials.

So, if the problem isn't going away, why don't the water utilities just set strict regulations for restaurants to ensure that they put the proper systems in place? Most utilities, with rare exceptions, are government-run and have the ability to both regulate and levy fines for noncompliance. Several of the utilities we reviewed do have such regulations. Yet, that alone does not solve the problem. Once the system is in place, it must be used consistently, requiring cooperation and vigilance by the restaurants. Unfortunately, this is not a high priority for many restaurants, especially the smaller ones, and the sewer blockages continue.

Our Mobile Society Equals Less Movement in the Sewers

For most cities, including San Francisco, the FOG problem of fats was not getting any better. In fact, things seemed to worsen as we became a more mobile society, increasingly dependent on the convenience of restaurants. And, even when we eat at home, there is a good chance that it will be take-out food. From our review of a number of utilities around the world, the story is pretty much the same: In New York City, its water utility cites the millions spent in clearing blocked sewers. In Dublin, the water utility's brochure on FOG shows pictures of large, boulder-sized solidified hunks of grease that had to be jack-hammered out of its sewers. In San Francisco, the annual cost for clearing its 2,500 blocked sewers hovered close to $4 million dollars, not including the costs or damages from sewer overflows.

"Cat-and-Mouse" Game between Utility and Restaurants

By 2006, San Francisco's water utility was running out of options in its battle with blocked sewers, sewer overflows, and all the resulting public health and environmental problems. Lewis Harrison, an environmental engineer with the utility's wastewater division, was the man in charge of making sure that sewer pipes (the sewage collection system) were operating properly. For several years, he had been trying different tactics to tackle the utility's FOG problem, including public education efforts, increased fines, and various other enforcement actions. They resulted in only minimal improvement in cooperation from the restaurant owners. And because the tourist industry is a powerful political force in the city—generating approximately 40 percent of the city's revenues—there was a limit to how much the restaurant fines could be increased.

Harrison and his staff found that many restaurants, especially the smaller ones, did not want to pay for grease removal. During inspections, the utility staff routinely found containers of oil stacked in the restaurants' basements and backyards, a situation that was not well-received by the city's public health department.

As another avoidance tactic, some of the restaurants began sending containers of grease home with kitchen clean-up staff. And, some restaurants, to avoid being fined, dumped their FOG not into the restaurants' drains, but instead directly into the city's storm drains. The result was the same: the grease caused sewer blockages. For Harrison and his colleagues, enforcement of the city regulations was nearly impossible. It had become a "cat-and-mouse" game between the restaurants and the utility; a game the utility was losing badly.

Waste Not, Want Not: From the Frying Pan to the Fuel Pump

Harrison, seeing that he was getting nowhere using conventional approaches, decided to try something very different: The city would no longer treat FOG as waste. Instead it would treat it as a commodity, an asset to be collected rather than discarded. He proposed to the utility management that the utility take the FOG and turn it into biodiesel, realizing that a portion of the FOG was a ready-made candidate for that purpose. The FOG in restaurants and fast-food shops consists of two types of grease: yellow and brown grease. Yellow grease is, most commonly, the used oils that come out of deep-fat fryers. It is collected in cartons or barrels until it can be hauled away. Brown grease comes from pans and dishes, and after it goes down the drain, it usually ends up in a grease trap or grease interceptor—a plumbing device—that is meant to catch the grease and keep it out of the sewer system. It is called brown grease because it is contaminated often with water and bits of food. Brown grease is removed when grease haulers come and pump it out of grease interceptors.

Removing the Economic Disincentive

Harrison came up with a program that would take the yellow grease and turn it into biodiesel. A key component of his proposal was free grease pick-up at restaurants. He realized that free grease collection would remove the economic deterrent to restaurant cooperation. He convinced the utility's management that it would result in better restaurant cooperation, fewer sewer blockages, and therefore less money spent on sewer maintenance.

A few months after Harrison had made his fryer-to-fuel proposal—in a move that was completely unrelated to Harrison's proposal—the city's elected officials started a program to require that biodiesel replace diesel in city-owned vehicles. The source of biodiesel for the city's program was to be the oil derived from "virgin" oils from soybeans grown in the Midwest. Harrison knew that if he was successful in turning the FOG into biodiesel, he could replace the "virgin oil" with his yellow-grease biodiesel and use it to power the city's fleet. For a water agency, promoting recycled "home-grown" fuel over the soybean fuel was a natural solution. The utility's yellow-grease program reduced the amount of grease going into the sewers and it was replacing a fuel source (soybean biodiesel) that consumes a lot of water to produce.

While the fryer-to-fuel concept seemed like an instant winner, there were several hurdles: How would the utility get the FOG from the restaurants? What steps were required to turn the grease into biodiesel?

The first hurdle for Harrison was finding a way to actually convert the collected grease into biofuel. Fortunately, he was able to find biodiesel processors in the San Francisco Bay Area that would process the "yellow grease" into biodiesel. These local green businesses were happy to enter into an agreement with the utility because it gave them a reliable source of feedstock for their biodiesel fuel production.

The utility's second challenge was to publicize the fryer-to-fuel program—with its free grease collection—to 2,500 restaurants and other food extablishments. The utility held a press conference launching the program at a local restaurant featuring the first free grease pickup. With the utility's strong media outreach, the program launch was well-covered by the San Francisco newspapers as well as radio, television, and other digital media.

SF Greasecycle Is Launched

The utility launched the program, which it named "SF Greasecycle," in December 2007. The utility's communication efforts also included setting up a dedicated multilingual website (http://www.sfgreasecycle.org), on which restaurants could sign up for free FOG collection and also obtain information on best practices for managing grease and other waste.

SF Greasecycle was well received by San Francisco's restaurant trade association, which helped promote the program to its membership. "What excited restaurant owners was the idea that the French fries they served one night could power the Muni (the city's transit system) bus they rode to work the next week," said Laura Spanjian, the utility's assistant general manager. The utility also got outreach help from its sister agency, the Department of Public Health, which provided information about the grease program when making routine restaurant inspections.

The program was an immediate success on several fronts. At the time of this writing, close to half of the city's restaurants have signed up for the program; the utility is collecting about 30,000 gallons of yellow grease a month, and more restaurants continue to sign on. The private haulers that had

Thirty thousand gallons of grease is collected per month and turned into biodiesel fuel, replacing virgin fuels grown in the Midwest with "home-grown" fuel.

been charging for collection and disposal of yellow grease are now offering free pick-up from restaurants.

The SF Greasecycle program has produced a number of positive results for the utility. Most promising is that the locations of restaurants requesting SF Greasecycle collection clearly coincide with locations of previous blockages and overflows in the city's sewers. And many of the restaurants signing up for the free collection are the smaller to mid-sized ones that previously did not have grease collection services. Harrison and his colleagues believe that this is a good indication that the program will succeed in preventing future sewer overflows. They have begun to see improvements in the sewers. Harrison said they are fighting years, in fact, decades of FOG buildup in the sewers; it may take years before they eliminate all the FOG blockage.

The program has been a financial success as well. The city utility and the private haulers are able to sell the "yellow grease" to the biodiesel processor for a price greater than their collection costs.

And, as Harrison predicted, there is more cooperation between the utility and the restaurants. It's no longer a game of "cat and mouse"; the utility and the restaurants are now focused on how they can get the job done together.

Moving on to the Bigger Challenges

While the SF Greasecycle program was a success, Harrison and his colleagues believed that the utility could do even more to minimize the FOG problem. In early 2008, they began working on more complicated projects. One was a plan to extend the grease-collection program to include the city's residences; the other was to better manage the brown grease problem.

Harrison knew that the residential collection of yellow grease would probably produce much smaller returns, but he believed it was worth the effort. He estimated that close to half of the yellow grease that ended up in the sewers came from homes throughout the city. Since pick-up from every residence was not feasible, Harrison and his colleagues began their efforts with a low-key outreach program at the local Costco warehouse store during the Thanksgiving holiday. The Costco store was selected because they were selling turkey fryers that require several quarts of cooking oil. Customers were asked to bring their vats of used cooking oils to the store for disposal. As of this writing, the outreach has expanded to five locations. The residential drop-off locations and other collection information are available on SF Greasecycle's website, and while the program has produced much smaller results, it is slowly catching on. In 2009, the utility collected 3,000 gallons of yellow grease from residential sources, a small amount in comparison to what the city collects from restaurants, but still worthwhile for the utility. "On the residential side, education pays greater dividends than just collecting grease. This program helps build awareness of the problems caused by grease going into the sewers," opined the utility's communications director, Tyrone Jue.

THE CONVERSION OF brown grease into biodiesel presented an even tougher challenge for the utility. Brown grease is often a nasty mélange of leftover animal fat, pan drippings, and other FOG residue that has already slipped into a restaurant's grease trap or down into the sewer pipes. In the grease trap, it often is combined with sundry bits of waste from the kitchen, including bits of paper and bones. In the sewer, it is then combined with water and all types of debris—grit and just about everything else that ends up in the sewer—and if you have been into a sewer, you know that means *just about everything.*

There has never been a viable program to convert brown grease into biodiesel because brown grease, unlike yellow grease, is filled with many impurities, including grit and water, that make it difficult to process into fuel. But Harrison and his colleagues believe it is possible. In early 2008, they submitted a grant proposal to California Energy Commission outlining their plans to develop a model program for recovering and recycling brown grease into biodiesel. Later that year, the state energy commission awarded the utility a $1 million dollar grant. It was also awarded a $200,000 grant from the federal EPA. Armed with $1.2 million, Harrison and his colleagues set about trying to grapple with the more complex problem of brown grease.

Through 2008 and much of 2009, they looked into finding an efficient conversion process. They approached a number of private contractors that had claimed to be able to take this mélange of grease-trap and sewer waste and turn it into biofuel. In late 2009, after searching for several months, they awarded a contract to a private vendor that helped the utility set up a test site at its Oceanside wastewater treatment plant. Since then, Harrison and his colleagues have successfully passed test runs to convert a troublesome substance of little value, brown grease, into a valuable commodity, biodiesel.

For Harrison's boss, Tommy Moala, the utility's assistant general manager and head of its wastewater division, the successful test runs spelled a major breakthrough that could offer promise to utilities throughout the world. "We thought we were cutting-edge [with] our yellow grease program, but being able to turn this slop [brown grease] into biofuel is a real breakthrough," said Moala.

For San Francisco's water utility, the SF Greasecycle program has been a great success and a model of what government and communities can accomplish when they have the will.

CREATIVITY AT THE SEWAGE TREATMENT PLANT: REDUCING ITS ENERGY USE AND GREENHOUSE GASES

Every week, 1.5 million pounds of animal blood flows forth from the Foster Farms chicken processing plants in Livingston, California. This deluge of blood and waste poses an enormous disposal problem, not just for Foster Farms, but for all poultry and meat processors. Such waste, if not properly treated, can easily seep into lakes, rivers,

East Bay MUD is the first water utility in the nation to convert meat-processing and household food waste into energy.

and underground aquifers, causing significant pollution to these water sources. Our next case shows how Foster Farms addressed their problem by forming an innovative partnership with a water agency located 100 miles to the north in Oakland, California. Their partnership helped Foster Farms dispose of potentially polluting organic matter, while producing renewable energy and reducing potent greenhouse gases.

East Bay Municipal Utility District, known to its customers as "East Bay MUD," is a water and wastewater utility that serves 650,000 residential and 15,000 commercial customers. Its service area on the eastern side of the San Francisco Bay includes the cities of Oakland and Berkeley. In 2000, the utility found itself with excess capacity in its wastewater treatment plant after several of its large waste-producing commercial customers from the fruit and vegetable canning industry closed their businesses. The utility management was faced with the possibility of raising the rates of existing customers to cover the ongoing costs of its wastewater facilities. So, the utility management spread the word to businesses outside its service area that it was willing to accept and treat their industrial waste.

In 2001, the utility started a partnership with Foster Farms, the largest poultry processors in the state, to process its chicken blood and offal. For all meat processors, this type of organic waste presents a significant disposal problem. The waste generated by the meat processors contains a high content of organic matter (blood, feathers, and fat) as well as nutrients like nitrogen and phosphorous. It can't just be dumped into a landfill or sewer system; if this type of waste is not treated properly, it can seep into groundwater or nearby streams and seriously pollute drinking water sources. In one noteworthy case, a chicken processor plead guilty in 2003 to 20 felony counts after not properly treating wastewater from its plants that wound up in a tributary of the Laramie River in Missouri. Though this case happened ten years ago, the problem has not gone away. In recent years, the EPA has cited several other meat processors for similar violations. In a 2009 case, a beef and pork processor agreed to pay $2.1 million in penalties for failing to properly treat wastewater discharges. The meat processor was discharging an average of one million gallons a day of effluent from its Nebraska plant into the Missouri River. The effluent contained fecal coliform and high levels of nitrates that harmed the river's aquatic life.

Blood into Energy

Foster Farms was "looking for a more sustainable method of disposal," said Jim Marnatti, environmental affairs manager of Foster Farms. So began the collaboration between the poultry processor and East Bay MUD. Foster Farms sends its animal waste via trucks to East Bay MUD,

Every week, 1.5 million pounds of animal blood flows to the East Bay MUD wastewater treatment plant.

where it is run through the utility's wastewater treatment process. This process includes several steps, the key component being the passage of animal waste through anaerobic digesters.

In wastewater treatment plants, anaerobic digesters can be best described as large concrete containers that function literally like a stomach. (See Figure 1.2 on pages 16–17.) In the digester, bacteria breaks down—essentially "digests"—the waste, releasing methane gas as one of its by-products. East Bay MUD captures the methane gas and then uses it as renewable energy to power its treatment plant. The solid waste that remains at the end of the treatment process can be used as compost for commercial applications.

Food into Energy

Motivated by its success in turning animal waste into renewable energy, and in an ongoing effort to further reduce energy costs, East Bay

"Who would have thought that food scraps would be powering your lights?"

—Cara Peck, Environmental Scientist, EPA

MUD decided to expand its organic waste disposal program. In 2006, with a grant from the EPA, the utility began to investigate the possibility of treating one of the least-recovered materials from the solid-waste stream: food waste. The utility tested the viability of processing ordinary food waste from restaurants, grocery stores, and food-service businesses, by sending it through the anaerobic digester—and turning it into energy. The initial results indicated that food waste could be successfully processed using the anaerobic digester system. In fact, the tests showed that food waste could be processed three times faster than regular sewage. The utility also found that the food waste produced three times as much methane gas on a per ton basis than did regular sewage solids. Quicker digestion and more methane gas production translated to a direct increase of its renewable energy output.

In 2007, based on its successful tests, East Bay MUD decided to invest $5 million in special pipes and screening equipment that would facilitate the new process. The utility then contacted garbage haulers in the San Francisco Bay Area and asked them if they were willing to send their food waste to East Bay MUD. Some responded positively, and, by early 2010, they were converting close to 40 tons daily of food waste into energy.

> *Food waste is the most difficult to divert from the landfills. Now East Bay MUD is turning 40 tons a day of food waste into energy.*

How big a deal was it for East Bay MUD to expand its operations to include the processing of food waste? It was a really big deal. The second largest portion of our waste footprint is food waste. In the United States, we waste close to 40 percent, or nearly 100 billion pounds per year, of what is available for consumption. We throw out everything from a blemished piece of fruit to a half-eaten sandwich. We are not alone: The amount of food wasted in Japan and the United Kingdom is also in the range of 30 to 40 percent.

> *In the United States, United Kingdom, and Japan, each wastes between 30 and 40 percent of their food.*

And, while food waste represents the second largest component of municipal waste, it is also the most difficult waste to divert away from landfills. Less than 3 percent is actually diverted. Yet, diverting food waste is important because food waste is a major contributor to methane gas buildup. Methane gas is 23 times more damaging to the environment than carbon dioxide, which is why landfills are one of the most potent sources of greenhouse gases.

> *Methane gas is 23 times more damaging in its greenhouse effect than carbon dioxide. A major source of methane gas is the food waste in landfills.*

East Bay MUD considers its treatment of meat processing and food waste as a successful venture for the utility. Its $5 million investment in equipment and retrofitting to facilitate food waste digestion will be paid back in just a few years, and the utility now produces close to 6 megawatts of electricity from its waste-digestion processes, more than enough to power its treatment plant; it is now on the verge of being able to send excess power back into the region's electrical grid. In the words of Ed McCormick, the utility's manager of wastewater engineering, "I never thought that a sewage treatment plant would be seen as a green factory."

East Bay MUD plans to expand its food-waste processing capacity from the current level of 40 tons per day to more than triple that amount. It is very likely that East Bay MUD will hit that goal because the San Francisco-based garbage-hauling and recycling company, Recology, Inc., has one of the Bay Area's most aggressive food-waste separation and collection programs. Recology collects an average of 300 tons of food waste per day, and it is negotiating with East Bay MUD to receive more of its food waste for processing into energy and soil compost.

The production of renewable energy, coupled with the reduction of potent greenhouse gases, are strong arguments for turning food waste into fuel. Every wastewater treatment plant with an

In California, there are 137 wastewater treatment plants with anaerobic digesters. These plants have an estimated 15 to 30 percent excess capacity. This excess capacity could be used to process food waste.

anaerobic digester should consider following the method employed by East Bay MUD to reduce its energy use—wastewater treatment uses 2 to 3 percent of our nation's energy—and to reduce a potent source of greenhouse gases. There is definitely more than enough food waste to power every treatment plant. Since water utilities are suffering the effects of climate change, it is important that they take a lead in combating it.

"This is a great opportunity, especially since our primary focus is public health and environment," said David Williams, East Bay MUD's head of wastewater. "Right now, we take a lot of carbon out of the ground and put it into the air. In this case, you're taking carbon that's already here and getting the energy out of it. That's the great thing."

SANTA ROSA AND THE GEYSERS

Our next case takes place in the heart of California's wine country, Sonoma County. In this successful public-private partnership, the city of Santa Rosa took its wastewater and used it to revive California's largest renewable energy plant. Or, as the saying goes: "One man's meat is another man's poison."

Many Californians think of Santa Rosa as a small town. San Francisco, its larger neighbor 50 miles to the south, considers it part of its suburbs. Yet Santa Rosa has grown from a small Spanish and Mexican settlement in the early nineteenth century to being the fifth largest city in the San Francisco Bay Area today. According to the town's latest general plan, the city's current population (2009) of 161,000 is expected to increase to 195,000 by 2020.

A public-private partnership takes troublesome sewer water and turns it into California's largest producer of renewable energy—more energy than all the solar and wind energy combined.

Santa Rosa's newer residents are more affluent than the Santa Rosa "old-timers." Evidence of the demographic shift can be seen in the older neighborhoods of small homes that have been dwarfed by subdivisions of 3,000-square-foot homes built close together, in cheek-by-jowl fashion. The affluent residents support new businesses and restaurants that draw some of the best chefs from San Francisco. Santa Rosa even has its own "home-grown" celebrity chef, Guy Fieri, with his own TV show on the Food Network.

But Santa Rosa's growth has not been without growing pains: those larger homes and additional businesses need more water and produce more sewage, which brings us to Santa Rosa's sewage problem. Actually, Santa Rosa has had sewage-disposal problems since the 1880s. The town had traditionally discharged treated sewage into a lagoon, Laguna de Santa Rosa, that feeds into the Russian River.

Santa Rosa's water and sanitation agency—like many others throughout the United States—discharged their treated sewage into a nearby waterway. And although the Santa Rosa Utilities department

Located 15 miles from Santa Rosa is the southern end of the Russian River, a popular vacation spot for many people in the region, including many San Franciscans who go there to escape the city's summer fog. The Russian River also has some other valuable visitors: it is the spawning ground for the endangered Steelhead trout.

had treated its wastewater to comply with all state and federal sewage-treatment standards, the practice was unacceptable to the residents of the smaller towns along the Russian River as well as the recreational and environmental advocates. There had been complaints and lawsuits against the town for decades. The battle reached a fever pitch in 1986, when it came to the public's attention that in the previous year Santa Rosa had dumped over 750 million gallons of treated sewer discharge into the river, an all-time high for the city.

The residents of California's wine country have a reputation for being "mellow." But after the news spread, they weren't mellow; they were angry. City manager Ken Blanchard, the official face of Santa Rosa, got so many irate calls that he was forced to change his phone number, and he needed a police escort when he attended public meetings. The Santa Rosa City Council and the Sonoma County Board of Supervisors, also feeling the heat, passed legislation that directed the City of Santa Rosa Utilities department to stop putting wastewater into the river. But the environmental and recreation advocates felt these legislative measures were hollow promises. For the next 15 years, the situation remained contentious and the wastewater kept going into the river. There were ongoing lawsuits by environmental and outdoor-recreation advocates claiming that continuing discharge into the river was a violation of the federal Clean Water Act.

The water agency understood that it had to do something, but it wasn't at all clear what that something would be. "We didn't have a lot of options," said Dan Carlson, the chief engineer at the water agency. "If we kept the status quo, we would continue to get hammered but the other options were not going to be easy, and they were expensive." One idea, to build large sewage-storage ponds in Santa Rosa, was met with fierce "not in my backyard" opposition. Another possibility, building a large outfall pipe 30 miles long that would discharge sewage into the ocean, faced environmental opposition and was also very expensive. A third possibility involved sending Santa Rosa's wastewater to the Geysers, a large geothermal power plant located to the north of Santa Rosa, which needed and was looking for a freshwater source to renew its energy production. Building the requisite pipeline to carry the sewage to the geothermal plant would also be expensive and presented several engineering challenges. For Carlson and his colleagues at the water agency, the prospects were not good. "For me, taking sewage to the Geysers was a real long shot. They were over 40 miles away and we were going to have to climb 3,000 feet up the mountains and cross a [earthquake] fault line; it didn't look good," recalled chief

Figure 6.2   Steam from a geothermal energy plant in California, now the largest source of renewable energy in the state. The energy is always available as opposed to wind or solar, which provide intermittent power. (*U.S. Geological Survey*)

engineer Carlson. A key difference, however, between the Geysers option and the others was that the Geysers needed the water and was willing to take Santa Rosa's wastewater.

The Geysers are natural hot springs, located in the Mayacamas Mountains about 72 miles north of San Francisco, that had become one of the country's largest sources of geothermal energy. The plant could produce more energy than all of California's solar and wind farms combined. The Geysers' geothermal energy has another advantage over wind and solar: The plant produces energy around the clock and is neither dependent on the sun nor the wind to produce that energy.

Water is key to the geothermal energy production process: Water is injected into the hot subterranean rocks, which produces steam that spins electricity-generating turbines.

In 1921, the first geothermal energy in the United States was generated at the Geysers and was used for heating and lighting at a nearby resort. In 1960, the Geysers began the nation's first commercial production of geothermal energy, producing 11 megawatts of electricity, enough to power 100,000 homes. The Geysers was operated originally by several power companies and then primarily by Pacific Gas and Electric Company (PG&E), northern and central California's large gas and electric utility made infamous by the movie *Erin Brockovich*. The Geysers reached the height of its production in 1987, producing 2,000 megawatts of electricity, enough to power over a million homes. Unfortunately, PG&E had oversized many of the steam plants, drying up the springs' natural water supply, and as a result, energy production at the plant decreased significantly. Earl Holley, an engineer working at the Geysers for 23 years, summed up the situation this way: "Basically, they put too many straws into the underground (hot springs) basin."

In 1989, the Calpine Power Company took over the operation of the Geysers from PG&E with plans to revive the energy plant. To do this it needed water, and lots of it. The company found a water source in nearby Lake County, some 11 miles away from the Geysers, but it would not be enough to bring the plant to its previous capacity. The Geysers needed more water and from a constant source. Although some of the water used in the geothermal process is recycled back into the system, most of it—about 80 percent—is lost through evaporation as it turns into steam.

While Calpine was looking for more water, Santa Rosa was still struggling to find a way to get rid of its ever-increasing sewer discharge. After five years of environmental analysis, studies, and endless public hearings, Santa Rosa selected shipping the sewer water to the Geysers as the preferred option. The partnership struck between Calpine and the city, which became known as the Santa Rosa Geysers Recharge Project, was a 30-year commitment that would both ensure the longevity of energy production at the plant and provide an environmentally sound means of using sewer water.

As Dan Carlson had predicted, this was a significant undertaking for Santa Rosa. It took $200 million, 700 workers, and three years to complete the pipeline. To no one's surprise, there were several complications, which included the political challenges involved in running a large wastewater pipeline through the wine country. Many of the businesses, residents, farmers, and property owners affected by construction activities complained, and several sought legal remedies. A total of twelve lawsuits were filed during that three-year period, but Santa Rosa's Utilities department prevailed in court.

The biggest challenge facing the project was "Mother Nature." The four-foot-diameter pipeline had to be constructed to withstand earthquake damage since it would cross fault lines similar to those that caused the deadly earthquakes throughout the San Francisco and the Bay Area in 1906 and 1989. It was more than a technical challenge; it was also a public relations one: Residents who were upset about construction activities near their property would surely be upset with a ruptured sewage pipeline. The pipeline also had to be constructed to scale a 3,000-foot climb up the Mayacamas Mountains. And the vertical climb was only part of the problem; the pipeline also had to be threaded through the Audubon Sanctuary, which involved a delicate dance by a massive pipeline through an environmentally sensitive area.

Thanks to the innovative Geysers Recharge Project, 19.8 million gallons a day of Santa Rosa's sewage no longer goes into the Russian River; it is providing electricity to close to one million homes.

Santa Rosa was assisted in its efforts by its new partner, Calpine. Despite struggling with bankruptcy, Calpine made a significant investment in the project. They spent $45 million on infrastructure improvements and provided the electric power that Santa Rosa needed to pump the wastewater through the pipeline to the Geysers.

The plan worked. With the new influx of water, Calpine was able to restore enough energy production at the plant to power 725,000 homes. By 2008, the system was working so well for Santa Rosa and the Geysers that they agreed to expand the flow of wastewater to the Geysers from the original delivery amount of 11 million gallons per day to 19.8 million gallons per day. With this expansion, the Geysers energy production will come close to providing power to one million homes.

The Geysers Recharge Project enabled Santa Rosa to end its 120-year-old wastewater problem. Spring 2009 marked the first year that Santa Rosa did not discharge sewage into the Russian River. Santa Rosa's then Mayor Bob Blanchard stated his support for expansion of the project: "This is another win-win for Calpine, for the City and our ratepayers. How can we get any more environmentally sustainable than to reuse our treated wastewater to help create electricity?" It was a wonderful turnaround from the days when City Manager Ken Blanchard needed police protection from an angry mob. This case is testament to perseverance by a government water utility, a struggling power utility, and a creative public-private partnership.

CHAPTER SEVEN

RIVERS AS
SHARED RESOURCES

Transboundary Conflicts and Compromises

Peter Gleick, president of the environmental research group the Pacific Institute, traces the history of water conflicts from 3000 BC until AD 2009. These disputes have ranged from minor to major and have sometimes resulted in skirmishes and wars, but since the 1940s there have been fewer violent conflicts and more attempts to resolve water issues by negotiation. But such negotiations, which may go on for years only to remain unresolved or to result in unsigned agreements, have had mixed results. It's not surprising why: the issues involved in such disputes are often enormously complex and of vital importance to the respective parties.

"There has been a lot of discussion about 'water wars,' a term that sounds great, but to which I do not subscribe: Wars start and are fought for many reasons, and while water has often been a target, tool, or objective of violence, it is certainly hard to ascribe the primary reason for any war to water alone."

—Peter Gleick, 2009

Lack of access to water can have major impact on the health and wealth of nations; major occupations, such as fishing and farming, cannot flourish, and the growth of cities is limited. With the development of nation-states in the sixteenth and seventeenth centuries, the lack of access by downstream users and the control of water by the upstream populace were firmly established. This meant that, without a treaty, the downstream users were essentially cut off from the use of the flowing river. The Industrial Revolution brought serious pollution to the rivers, which also impacted the downstream users.

Over time, several legal principles or doctrines have been invoked to help resolve the potential for conflict between nations sharing a river water source. For example, upstream countries often claim *absolute sovereignty*. In other words, they take the position that they cannot be restrained in their water use by any other country in the river basin. For example, Turkey is upstream of its neighbors Syria and Iraq on both the Tigris and the Euphrates Rivers. Under the doctrine of absolute sovereignty the upstream country has every right to control the water as long as it is flowing within its own national boundary. This does not mean that the upstream countries cannot be friendly with their downstream neighbors or work out a reasonable outcome, but it does rely upon the good neighborliness of the upstream peoples. In the case of Turkey and its neighbors in the Tigris and Euphrates river basins, Turkey has over the past 20 years built 12 storage reservoirs in those parts of the river basins that lie within Turkish national boundaries. The projects, called in total the Greater Anatolia Project (GAP), when completed—ten more dams are planned—will store and control more than 50 percent of the natural flows that would have crossed the borders into Syria and Iraq if the projects had not been built.

A legal principle beloved of the downstream countries is *riparian rights*. This is the idea that all the countries that lie in the river basin and along the river (the riparians) are entitled to an equitable share in its water. According to the downstream countries, adherence to the riparian rights doctrine would make sharing among upstream and downstream countries much more equitable than the absolute sovereignty principle discussed above. The riparian rights principle, which is based on Roman law, is embedded in the water laws of many European countries. It is also espoused by Bangladesh and many other countries that lie downstream on both the Ganges and the Brahmaputra rivers. These downstream countries have to rely upon the goodwill of India, China,

Nepal, and Bhutan for access to water during the low-flow seasons and suffer from excessive flooding during the monsoon season due to flood control works in the upstream countries that increase the amount of water flowing downstream. Indeed, India and Bangladesh signed the Farakka Treaty in 1975 in an attempt to deal with just the low-flow part of the problem caused by diversions of flow upstream of the Bangladesh border to provide more water flow in the port of Calcutta. The Bangladesh government still believes that it did not get its fair share of the water, and this remains a major source of friction between the two countries.

One other doctrine, much beloved by countries that have long histories of water development, regardless of their location in the river basin, has to do with when the water was first appropriated. This is called *prior appropriation—first in time, first in right* and it was first made popular in California and other U.S. western states during the Gold Rush, when miners needed large quantities of water for sluicing the gold out of the sediments. Since water was not necessarily located where the best deposits were, miners simply took water (appropriated it) from the nearest stream. "First come, first served" was the rule of the day and woe to any latecomer since water disputes were often settled using Samuel Colt's "peacemaker" revolver. Over time, this principle has become codified in the water laws of the western states, while in the eastern states water laws are typically based on riparian rights. Egypt is also a firm supporter of first in time. Despite being the most downstream country in the ten-nation Nile Basin, Egypt claims prior appropriation based on its 4,000-year history of near-exclusive diversion of the Nile waters for irrigation. And Egypt still claims about 70 percent of the Nile's total flow even though it essentially makes no contribution to the river flow in terms of rainfall. This is a major source of unhappiness in the upstream countries of the basin.

The UN's International Law Commission spent 21 years, from 1971 to 1992, drafting the UN *Convention on the Protection and Use of Transboundary Watercourses and International Lakes*, which was largely based upon the widely invoked principles of *prior consultation; avoidance of significant harm; equitable apportionment;* and the *provision for settlement of disputes*. These common-sense principles ensure that countries inform their neighbors of their intentions (prior consultation) to disrupt the flow of transboundary rivers; that they try to avoid causing significant damage (avoidance of significant harm) to the economies and the ecosystems of the downstream riparians; that every attempt should be made to ensure equitable apportionment of the disputed waters or the economic benefits or damages resulting from upstream countries' actions; and that there ought to be a venue (provision for settlement of disputes) where irreconcilable disputes can be resolved. Although it has not yet been ratified in the UN General Assembly by the requisite 35 countries, the existence of such a treaty is a good indication of the international community's intentions to improve collaboration among riparians. But the failure to ratify it also says a lot about the desires of upstream countries not to cede sovereignty to a supranational body. Moreover, the Convention lacks an effective enforcement mechanism; thus, even when it is ratified, carrying it out will still require the goodwill of upstream countries, or the hegemonic strength of the downstream countries. In any case, the fact that it has not yet come into force has not hindered the resolution of many smaller water conflicts using the treaty's common-sense ideas.

At the 2008 World Economic Forum in Davos, Switzerland, UN Secretary General Ban Ki-Moon implicated drought as a cause of the ethnic conflict in Darfur. Cautioning that a shortage of water resources could cause similar conflicts elsewhere, he pledged UN action, noting that "Our experiences tell us that environmental stress, due to lack of water, may lead to conflict, and would be greater in poor nations. . . . This is not an issue of rich or poor, north or south." Moon illustrated his claim with examples of transboundary water problems in China, the United States, Spain, India, Pakistan, Bangladesh, and the Republic of Korea.

Water conflicts seem to arise every time a river crosses a boundary. For instance, in the Colorado River Basin, despite the existence of the 1928 federal law that apportioned the water of the Colorado River, there are still serious water disputes about the quantities of water used by individual states among the seven U.S. basin states (Colorado, New Mexico, Utah, Wyoming, Nevada, Arizona, and California) and Mexico. In India, we see similar conflicts regarding the Ganges River, both domestically and internationally, with Nepal and Bangladesh sharing many rivers with India. Despite several bilateral agreements between India and Nepal and India and Bangladesh, there is no overall treaty that binds the development of the river into one coherent whole.

In the following sections, we report on attempts to mitigate conflicts involving three international rivers: the Indus, the Nile, and the Mekong (known as the Langcang in China). The efforts for the Indus River were quite successful, while the difficulties with Egypt and China underscore the complexity of the problems of the Nile and the Mekong.

> *Globally, over 261 rivers cross two or more international boundaries.*

TRANSBOUNDARY RIVER BASINS

Because of the mounting pressure on water resources due to increasing population and wealth, attention has begun to focus on the efficiency of water use, and disputes about the use of water from transboundary rivers has become more contentious. Along with this increased pressure, we are only recently discovering the wide scope and sheer magnitude of the problems associated with managing transboundary rivers. Experts estimate that there are over 145 countries participating in one or more of the 261 international river basins on Earth. There have been as many as 300 river-sharing agreements in

Europe since the Treaty of Versailles in 1815, which oversaw the re-arrangement in Europe after the Napoleonic Wars. However, almost all of them regulated in-stream use for navigation, hydropower, fishing, and pollution disposal and did not involve the type of large-scale diversions of water that are now common in irrigation development. Large withdrawals of water typically create very difficult water-allocation problems for the downstream countries.

A period of great concern about transboundary river conflicts was in the 1950s and early 1960s when decolonization was sweeping the globe and newly independent countries began to experience the problems of river management. This period culminated in 1960 in the Indus Waters Treaty between India and Pakistan, brokered by the World Bank. The accord fueled optimism that other major water conflicts could also be resolved. At the time, basins such as the Ganges-Brahmaputra, the Mekong, and the Nile (and even the tiny Jordan River basin) were subject to detailed development planning, and river-basin commissions were created by the UN in an attempt to avoid conflict among the users. Sadly, as we will see in the Nile and Mekong cases below, most of these efforts came up short. Currently, we are experiencing a renewal of disputes fueled by the water shortages caused by rapid development and huge population growth, and possibly climate change.

Decisions about how to allocate or reallocate the flows of a river always involve politics. No matter how detailed the technical, economic, and social studies are, hard choices have to be made among the various users who stand to gain or lose from such accords. This is true whether the river is a national river or crosses international borders. However, disputes over transboundary rivers require a level of political decision making that goes beyond local and national interest groups. It requires the ability and the willingness of sovereign nations to negotiate.

All rational planners recognize the value of cooperation on river-sharing issues, whether socio-cultural or trade and economic ones. What is not clear, however, is how to put a value on cooperation.

River Basin Politics

The problem with purely political decisions is the lack of predictive behavior on which they are based. Thus, many political decisions have looked beyond the scientific-technical analysis to take into account the qualitative benefits of resource sharing among coalitions. Political considerations are sometimes heavily influenced by noneconomic factors, such as national pride or willingness to reward different social groups with varying payoffs, and are often pursued separately and apart from economic objectives. Ultimately, foreign policy is the most influential determinant of a country's position on international rivers. It is important to link the potential river settlement to other pending economic and social issues between and among countries.

A wide range of solutions is possible in most water negotiations, but the net benefits are not the only consideration; a lot of political issues may dominate, such as which interest group is being served. These political considerations are not necessarily compatible with sound water allocation.

SOURCE OF CONFLICTS

We saw in Chapter 5 that managing common-property resources is a very difficult endeavor; managing transboundary water resources is no exception. River flows have both negative and positive externalities typically working only in one direction, that is, downstream. This pervasive unidirectional feature of water use generally rules out the resolution of basin conflicts through mutual control of external effects

> *"An international river is a common property resource shared among the basin states."*
> —Le Marquand, D. G.
> *(1977)*

based on conditions of reciprocity. However, downstream countries can also benefit from some positive external effects of upstream use. Aside from the water-allocation problems that arise from the physical sharing of a common resource, there are also many water-quality problems that can arise downstream as an effect of upstream use. Table 7.1 lists some of the most common physical externalities that result from water use.

TABLE 7.1   DOWNSTREAM EFFECTS OF UPSTREAM WATER USE. FROM ROGERS (1993)

WATER USE	EFFECTS
Hydropower	
Base load	Helps regulate river (+ve)
Peak load	Creates additional peaks (-ve)
Irrigation diversions	Removes water from system (-ve)
Flood Storage	Reduces seasonal flows in the river (-ve)
Municipal and Industrial	
WS diversions	Removes water from the river (-ve)
Wastewater treatment	Adds pollution to river (-ve)
Navigation	Keeps water in river (+ve)
Recreational storage	Keeps water out of the system (-ve)
Ecological maintenance	Keeps low flows in river (+ve)
Groundwater development	Reduces groundwater availability and base flows (-ve)

INDIRECT USE

Agriculture	Adds sediment and agricultural chemicals (-ve)
Forestry	Adds sediment and chemicals, increases peak runoff (-ve)
Animal husbandry	Adds sediments and nutrients (-ve)
Filling wetlands	Reduces ecological carrying capacity, increases floods (-ve)
Urban development	Induces flooding, adds pollutants (-ve)
Mineral deposits	Add chemicals to surface and groundwater (-ve)

The plus signs in the table indicate positive downstream effects and the minus signs indicate classical downstream damages associated with upstream development. Natural processes such as floods and droughts can also cause major downstream effects and are sometimes mistaken for manmade externalities, leading to further mistrust and increased tension among the riparian states.

WHAT MAKES A SUCCESSFUL TREATY: THE CASE OF THE INDUS RIVER

Concerning the signing of the Indus Waters Treaty by the prime minister of India, Jawaharlal Nehru, and the president of Pakistan, Ayub Khan, on September 19, 1960, Sir W. A. B. Iliff, the World Bank Representative, said:

> The Indus Waters Treaty was the outcome of eight years of discussion and negotiation between the Governments of India and Pakistan, carried on under the auspices of the World Bank. It brought to an end the long-standing dispute between India and Pakistan about the use, for irrigation and hydro-power, of the waters of the Indus system of rivers.

The Indus Basin, prior to the partition of the former territory of India into Pakistan and modern India in 1947, was the oldest and largest

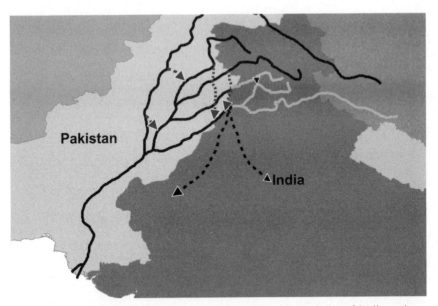

Figure 7.1 Map of Indus Basin showing the national boundaries of India and Pakistan. The dotted lines show the major river diversions under the Indus Treaty.

integrated irrigation system in the world. The irrigation works were located in what was called the Punjab (five rivers), for the five major tributary rivers of the Indus that ran through the basin: the Jhelum, the Chenab, the Ravi, the Beas, and the Sutlej. After partition, most of the water-abundant headwaters of the river system went to India, and Pakistan was the lower-basin riparian (see Figure 7.1). This was ironic since 80 percent of the irrigated area was actually left in the downstream country, Pakistan. The dispute between India and Pakistan over Pakistan's access to irrigation water came to a head in March 1948 when India cut off water supplies to some of the canals, affecting over 55 percent of Pakistan's irrigated area. A temporary agreement (the Inter-

> World Bank involvement did not come about easily; in fact in 1950, India's Prime Minister Nehru rejected any suggestion of third-party adjudication as being "a confession of our continued dependence on others."

Dominion Accord) was reached on May 4, which was only valid until October 1948. There followed three years of argument and bluster between the parties, until the World Bank was invited to help mediate the conflict.

The participation of the World Bank was agreed to eventually by both parties because each had tried to block a World Bank-funded project in each other's territory. So, India objected to the Bank's financing of the Kotri Barrage, a dam on the Indus within Pakistan, and Pakistan objected to the Bank's proposed funding of the Bhakra Dam on the Sutlej River in India. Since the Bank was unwilling to alienate either country, this impasse could only be bridged by agreeing to use the Bank as a third-party adjudicator.

In 1952, World Bank president Eugene Black offered the World Bank's assistance in attempting to resolve the dispute. In doing this, Black was supported by close friend David Lilienthal, former chairman of the Tennessee Valley Authority (TVA) and of the U.S. Atomic Energy Commission, who had implemented the TVA, the cornerstone of President Roosevelt's successful social experiment of using river basin development to trigger economic and social development during the New Deal. This initiated what would turn out to be a more than

30-year-long undertaking: eight years of negotiation with a treaty signed in 1960, followed by 22 years of construction to implement the agreements.

In 1953, both countries put forward plans for the basin that were irreconcilable; essentially, each side proposed to use all of the water. The Bank's representative put forward its plan, which relied upon meeting historical withdrawals not necessarily from the same source; the three eastern rivers were assigned to India, and the three western rivers were assigned to Pakistan; and Pakistan would have five years to adjust to the replacement of water supplies from India.

By 1954, the two countries had agreed to the basic engineering solutions, but it took another six years to iron out the political sovereignty issues over the sources of water in the agreements. Moreover, by 1959 it was apparent that the economic costs of the proposed investments needed to make the treaty function were beyond what the countries could be expected to pay and what the World Bank could afford to loan out. The Bank's panel of experts, however, found that the proposed plan would not work without building storages in each part of the basin. The final cost was almost US$1 billion in 1960 (about $7 billion at current values). Not until 1960 was the World Bank able to ensure both countries that sufficient aid funds would be forthcoming from donors (United States, Canada, United Kingdom, Germany, Australia, and New Zealand) to finance the investment requirements of the treaty.

The fact that The Indus Waters Treaty between two powerful adversaries has lasted 50 years and that the Indus Basin Commission continued to operate during two wars between India and Pakistan is a testament to its success. In essence, there were five conditions that led to the successful outcome:

- innovative technical proposals;
- willingness of both countries to compromise on political issues;
- the participation of powerful political personalities in both countries, such as Jawaharlal Nehru, the prime minister of India, and Ayub Khan, the president of Pakistan;
- the involvement also of powerful political actors in the donor countries, such as David Lilienthal in the United States; the World Bank as a powerful arbitrator;
- and obtaining the necessary financing.

Recently, because the river basin has been fully developed for irrigation diversion uses, proposals to use the water without diversion for hydroelectric power have become a source of tension between the countries; nevertheless, given the longevity of the arrangement and the clever language of the treaty itself, these conflicts are unlikely to destroy the overall treaty mechanisms.

Looking toward the many current international water conflicts it is hard to see how the nations involved will be able to reproduce the climate for negotiation enjoyed by the Indus countries. Moreover, in many cases there are several nations involved, not just two. The Nile has ten, the Mekong has six, and the Tigris-Euphrates has three major and three minor participants. Without the five conditions experienced during the Indus negotiations' being in place, it will be difficult to make significant progress on the many pressing current transboundary river conflicts around the world.

NILE RIVER BASIN: TEN COUNTRIES IN SEARCH OF AN AGREEMENT

The 4,135-mile-long Nile River flows through ten countries in its long journey from southern Rwanda to its delta on the Mediterranean Sea. The basin has been settled for more than 4,000 years and is considered the cradle of civilization. Over the millennia, empires have risen and fallen, but the Nile has endured as a symbol of the impermanence of human institutions and the durability of natural systems. Since the dawn of history, irrigation along the banks of the Nile has thrived. Together with fishing, farming along the Nile River provides sustenance to 300 million people. In the past, the dominant country controlled the waters of the Nile—typically it was Egypt. Since 1959, Egypt and Sudan have essentially split the total water of the Nile between themselves, ignoring the other countries involved.

In 1999, the World Bank helped establish the Nile Basin Initiative (NBI) in an attempt to negotiate a water sharing agreement among the ten riparian countries: Egypt, Sudan, Ethiopia, Uganda, Kenya, Tanzania, Burundi, Rwanda, the Democratic Republic of Congo, as well as Eritrea (see Figure 7.2). The goals sought by the NBI are that the cooperation of the Nile riparians could be a catalyst for regional economic integration and also key to resolving regional issues, civil wars, and poor economic performance by the individual countries. It could also lead to possible liberalization of labor and capital markets, thus making water infrastructure projects a means to an end. The main question is: Can the benefits of regional economic integration be achieved by making optimal system-wide investments instead of suboptimal investments in water infrastructure in each, or several, of the countries?

It was hoped that cooperation could realize joint gains from rural electrification and economic development particularly upstream (Ethiopia and Uganda) and irrigation development downstream. It could preserve the region's environmental assets instead of sacrificing them for short-term economic gain, including the Sudd, the world's largest freshwater wetland and winter home to Europe's birds; the great canyons of Blue Nile gorge; the Lake Victoria freshwater ecosystem, and the forests of Ethiopia's highlands.

The Nile Basin Initiative's vision separates the question of who has property rights to the Nile waters from the question of where the water might best be used for irrigated agriculture and other purposes. Even if no binding riparian agreements on water rights were reached, it was

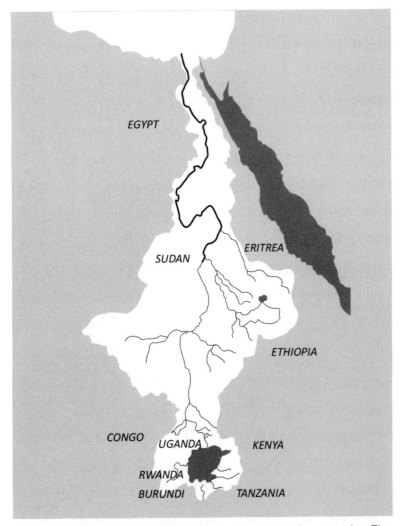

Figure 7.2   Map of the Nile Basin showing the ten riparian countries. The shaded area to the right is the Red Sea.

hoped that long-term leases could be made and could be renegotiated from time to time. A basin-wide authority could oversee the leasing, similar to the Murray-Darling Basin Authority, discussed earlier in Chapter 3.

After ten years of the Nile Basin Initiative, it was widely expected that such a commission would be created; however, despite the drafting of the Cooperative Framework Agreement (CFA) in May 2009, which would have created such a commission, opposition from the two large downstream countries, Egypt and Sudan, has delayed the signing of the treaty. These two countries claim historical rights based on two documents: the 1929 agreement between Egypt and Great Britain that granted Egypt the rights to most of the Nile's water flow and veto power over upstream projects, and the 1959 treaty between Egypt and Sudan, which effectively split the waters between them. The upstream countries claim that these agreements are invalid and that the status quo is unfair. While they (the other eight countries) are still hopeful for renegotiation of the CFA, the impasse underlines the difficulty in managing transboundary rivers when large and powerful nations refuse to collaborate. Nevertheless, despite the unhappiness over the failure of the CFA, the Nile Basin Initiative has already substantially increased investment in the basin. Over $1 billion in grants and loans has been spent on current and planned projects for hydroelectric power, data gathering, environmental protection, and agriculture.

THE MEKONG: IS 50 YEARS TOO LONG TO WAIT?

The Mekong River case provides a good example of a transboundary river dispute requiring international intervention. The Mekong flows for 2,900 miles, from Tibet to the coast of Vietnam. It is the world's twelfth longest river, and places in the top ten in terms of volume discharged. It passes through China, Laos, Thailand, Cambodia, and Vietnam, and borders on Burma (Myanmar). A fact of capital importance for its ecology and management is that 44 percent of its total length lies within China (see Figure 7.3).

In 1957 the Committee for Coordination and Investigation of the Lower Mekong Basin (the Mekong Committee) was established by the United Nations Economic Commission for Asia and the Far East (ECAFE), and it included all the lower basin riparians (Thailand, Cambodia, Laos, and Vietnam). It proceeded to generate very ambitious plans for the full-scale development of the basin. Four mainstream dams were planned for completion by 1987. However, by the time the committee was ready to implement them, the region was mired in the Vietnam War, and by 1973 the international community had lost much of its interest in helping develop these projects. An Interim Mekong Committee was reconstituted with only three of its original members under the auspices of the United Nations Development Program (UNDP) with headquarters in Bangkok. None of the proposed major projects were undertaken by the old committee or the revised interim

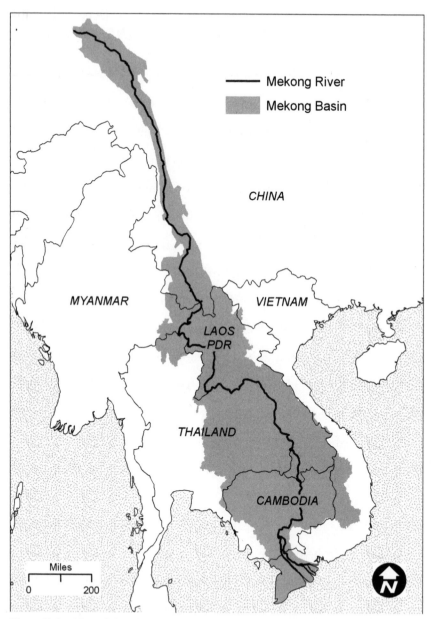

Figure 7.3   Map of the Mekong River Basin showing the national boundaries of the six riparian countries.

committee, and development was limited to small projects on the tributary rivers located within national boundaries. In 1995 the interim committee was replaced by the original four countries as the Mekong River Commission (MRC), and the institution was revived with renewed interest in developing large, jointly funded transboundary projects on the main river.

Since the 1980s the character of the river has been steadily transformed by China's unilateral dam-building program on the mainstream of the Mekong River, called the Langcang in Yunnan province. By 2009, three hydroelectric dams were already in operation, and two more large dams were under construction and due for completion in 2012 and 2017. Plans exist for at least two further dams, and by 2030 there could be a "cascade" of seven dams in Yunnan, all on the Langcang. China has acted without consulting its downstream neighbors, but because the current dams are relatively small and essentially involve very small amounts of over-season storage, their downstream impact so far has been limited. (Small dams cascading down a river valley producing hydroelectricity have small impacts on the flow regimes downstream.) However, future large upstream storages could cause serious damages to irrigation and fisheries downstream.

Moreover, despite the historical efforts, until recently there have been no firm construction plans for dams on the mainstream of the Mekong downstream of China. This situation has changed over the past three years when Memoranda of Understanding to build and operate projects between some of the countries have been signed for 11 proposed dams, including some on the main river that are being backed by private capital or Chinese state-backed firms. Particular attention and concern have been articulated by the NGO community about two sites: Don Sahong at the Khone Falls in southern Laos and Sambor in northeastern Cambodia. If built, these dams could block the fish migrations that are essential to ensure the food supplies of Laos and Cambodia. However, some NGOs claim that even small fluctuations in the hydrology of the Mekong could damage the agriculture, fisheries, and ecosystems of Laos, Thailand, Cambodia, and Vietnam, the countries of the Lower Mekong Basin (LMB).

For the LMB countries the Mekong is a major source of irrigation water and fisheries. In Vietnam's Mekong Delta the annual pattern of allowing flooding in the wet season and planting as the flood waters recede (flood recession agriculture) provides over 50 percent of agriculture's contribution to the country's gross domestic product. For all LMB countries, particularly their role in regulating Cambodia's Great Lake (Tonle Sap), which is a bountiful source of migratory fish, the natural flooding of Mekong water is essential. The annual value of the catch is US $2 billion, and more than 70 percent of the Cambodian animal protein supply comes from the river's fish. Eighty percent of the Mekong's fish species are migratory, some traveling many hundreds of miles between spawning and reaching adulthood. Eighty percent of the current population of the LMB depends on the Mekong River for sustenance, either in terms of wild fish captured in the river or through both large- and small-scale agriculture and horticulture.

Even without dams on the mainstream, the cascade of existing and soon to be started hydroelectric power plants could ultimately have serious effects on the functioning of the Mekong because altering the hydrology may cause the current "flood pulse" to be attenuated. This "pulse" causes the regular rise and fall of the river on an annual basis, which plays an essential part in the timing of spawning and the migration pattern of the fish. It could also block the flow of sediment downriver, which has a vital role both in depositing nutrients on the agricultural regions flooded by the river and as a trigger for fish migration—it is estimated that over 50 percent of the river's sediment comes from China. Ultimately, the sediment will be trapped behind the dams and not pass down to the lower agricultural reaches. Other economic, social, and environmental impacts could occur due to increasing the amount of flooding, most importantly in Cambodia and Vietnam, which could also lead to the erosion of river banks.

One would think that the existence of the Mekong River Commission (MRC) would have been the major line of defense against the threats posed by both the upstream Chinese dams and those now proposed for the downstream reaches. However, the MRC does not have the power to act to control these impacts. It essentially is a clearinghouse and "talking shop" where the individual plans can possibly be coordinated, but not mandated. It would be nice if there were a body able to mandate or control what individual countries choose to do on their sections of the Mekong, but that is wishful thinking in such a contested basin. The noninclusion of China and Burma in the 1995 agreement establishing the MRC underlines the body's weakness. The existence of the MRC has apparently had no effect on the national governments' pursuit of projects of purely national self-interest. The Mekong River riparians are now at a critical juncture: Most of the projects built or planned have been relatively small in terms of storage and diversions, and no major dams have yet been built on the mainstream of the river in the downstream reaches. There is still time to plan to mitigate the worst effects of the new developments as they come on line. However, if the countries do not cooperate, the future of the Mekong as a great source of food, both through fish and agriculture, could be in serious jeopardy.

Why has the MRC in the Mekong not been as successful as the Indus River Basin with its treaty and its subsequent evolution? In terms of the conditions identified in the Indus case for successful collaboration on a transboundary river, we can point to the following inhibitions to the success of the MRC. For example, the Indus had only two large and powerful countries in contention, while the Mekong has six countries with huge discrepancies in wealth, land areas, and population. In the MRC, planning is much more diffuse than the Indus: there are many more conflicted issues, not just irrigation flows, but sediment movement, fisheries development, and induced floods and water shortages in the downstream reaches; many of the borders are not contiguous; the countries have hugely different social and human cultures and a wide disparity of types of governments and political cultures; and there is a noted absence of third parties pushing for effective transboundary developments. These limitations do not necessarily mean that the MRC will not be able to build a stable coalition for the sustainable development of the river basin, but rather that it may have to work harder to convince all six of the riparians that it is in their best interest to collaborate on joint development and management of the river.

CHAPTER EIGHT

WATER THAT LASTS A THOUSAND YEARS

Bottled Water

When bottled water began appearing on the shelves of grocery stores and supermarkets in the 1970s, many people were amazed that customers were actually paying for something that you could get from a tap for free. Two decades later, we are even more amazed at how much we have to pay to clean up the environmental damage caused by all those plastic bottles. Increased awareness of the potential dangers of plastic, both in landfills and as a drinking vessel or container, has resulted in the beginnings of a mini-backlash. In Australia, a small town of 2,500 recently voted to outlaw the sale of bottled water. Would large cities in the United States embrace such a

Bottled-water sales worldwide are projected to reach $65.9 billion per year by 2012.

policy? Sad to say, there would probably be a major public outcry. Americans consume more bottled water than any other country in the world, spending close to $12 billion a year.

How did we get to this state of affairs? Since the seventies, when small, green bottles of Perrier became the must-have status symbol at dinner parties, bottled water has grown to become a permanent fixture in American life. While there are a number of reasons for this explosion, not all of them bad, it nevertheless has largely resulted from a combination of corporate profit-seeking, marketing, and advertising, and consumer gullibility. The profit margins for bottled water are significant—as one newspaper asked, "What's Colorless and Tasteless and Smells Like . . . Money?"—and we are easy advertising targets. To quote the late comedian George Carlin, "What is Evian spelled backwards?"

Apart from the costs involved in collecting, bottling, and distributing the water, it takes tons of fossil fuels to manufacture and distribute the plastic (or, sometimes, glass) bottles it comes in. After that, there is the expense of transporting the heaps of used bottles that end up in landfills. These are all good reasons to be opposed to bottled water.

Beyond cost and environmental concerns, our primary objection to bottled water is that bottlers' extensive marketing campaigns have helped erode the public's confidence in our own tap water. We have spent trillions of dollars establishing great water systems throughout this country, but it costs money to maintain them. These systems are in desperate need of infusions of cash. But we are more eager to spend money—$11.5 billion in the United States (2007), 1.5 billion pounds in the United Kingdom (2008)—on tap water

> *We are spending billions on bottled water, yet we hesitate to spend money on the water systems essential to our everyday well-being.*

packaged into bottles than to invest in the water systems themselves. Unless we are planning to bathe and cook with bottled water, our money is better spent on our own water and wastewater systems.

In this chapter, we describe several David versus Goliath attempts to champion lowly tap water over high-priced bottled water. We will also describe how one small town was able to avoid losing control of its water source to a bottled water company. And, finally, we discuss how we may have gotten too complacent to recognize how lucky we are with our ready access to safe drinking water.

Let's start with some basic information about bottled water:

- Worldwide consumption has steadily increased over the last thirty years. In 2008, demand started to flatten in North America and Europe, but overall sales have been bolstered by the increasing sales in China.

- The United States is the world's largest consumer of bottled water. Our annual consumption has grown from 1.6 gallons per person in 1976 to 28 gallons per person in 2008.

- The largest per capita user is the United Arab Emirates with 68 gallons per person.

- Less than 20 percent of plastic water bottles in the United States are actually recycled.

- Bottled water companies have fought recycling laws requiring deposits on bottles. In 2009, the industry fought such a recycling law in New York even though there were an estimated 3.2 billion water bottles sold in the state.

> *Americans discard 50 billion plastic water bottles a year; most of those bottles are not recycled.*

- In the United States, the fossil fuel consumed in one year in the production of plastic water bottles could fuel one million cars.

- A significant percentage of the bottled water that is sold in the United States comes from international sources, which means that most of it must be shipped thousands of miles—using lots of fossil fuel—to reach U.S. consumers. For example, Fiji-brand bottled water travels 7,950 miles by ship and truck to reach New York City.

- It is estimated that between 25 and 40 percent of bottled water comes from a public tap, but we can't be sure because most states and the federal government don't require that the bottled water companies disclose their water sources.

- The cost of bottled water is anywhere from 200 to 10,000 times the price of tap water.
- Bottled water that is sold in the same state in which it is bottled is not subject to federal inspection or regulation.

In summer 2009, the General Accounting Office released a report that revealed that Federal Food and Drug Administration rules don't require certified laboratories to conduct testing of bottled water nor publicly disclose any contaminants it may find. By contrast, tap water is regulated by the EPA, which requires continual testing by certified laboratories that must issue detailed quality reports; any detection of contaminants must be disclosed to the public. The cities of San Francisco and New York, for example, although their tap water comes from pristine sources, test it an average of 80,000 times over the course of a year and make results available to the public.

Over the past four years, there has been heightened awareness about the need to wean ourselves from the bottle. Some experts think this is reflected in the slight decrease in consumption in North America and Europe. Other experts view the decrease as a result of the economic recession. Our opinion is that recent attempts to raise public awareness to some of the bottled-water hype has prompted many consumers to think before they buy. Here are some of those attempts to increase awareness.

THE HIGH PRIESTESS OF CUISINE GETS OFF THE BOTTLE

In restaurants, bottled water is a big mark-up item. So, it is interesting to note that some San Francisco Bay Area restaurants that were quick to offer bottled water in the past are now offering tap water as their first choice. One of the key people behind this change in attitude may well be the high priestess of cuisine, Alice Waters. In 2006, Ms. Waters, the revered inspiration behind many innovations in American cooking and eating habits, took a strong stand against bottled water and replaced it with tap water in her restaurants. Her Berkeley, California, restaurant Chez Panisse had been selling about 25,000 bottles of water a year. The restaurant's general manager, Michael Kossa-Rienzi, put their decision in perspective: "All the energy to bottle water, carbonate, put [it] in glass, ship it and truck it to our restaurant; it was such a waste."

The changes at Chez Panisse inspired other restaurants to follow suit and may have been a factor in San Francisco's decision to limit spending public funds on bottled water. The city had hesitated for over two years to adopt a policy that would curb its spending on bottled water—some elected officials even advocated going into the bottled water business, that is, taking the city's tap and bottling it. In February

2007, the heads of San Francisco's Department of the Environment and its water utility wrote an op-ed piece in the *San Francisco Chronicle* arguing against the city's continued purchase of bottled water. Then Alice Waters came into San Francisco in March 2007 and used one of the city's public buildings (the historic Ferry Building) to make her announcement that Chez Panisse had forsaken bottled water. Within months, City Hall decreed that it would stop spending city money on bottled water.

Now the ball was rolling. The June 2007 United States Conference of Mayors, led by the mayors of Salt Lake City, San Francisco, and Albuquerque, among other cities, urged its membership to limit expenditure of public funds on bottled water. Today over 50 cities in the United States and Canada limit or prohibit the spending of public funds on bottled water.

What Happened to the Water Fountain?

Okay, so we want to follow Alice's advice and kick the bottled water habit, but where do we get easy access to tap water when we're on the go? In the old days, we might have headed for the nearest water fountain, strategically placed in schools, shopping centers, and other public places. Whatever happened to the water fountain? In the anecdotal information we have gathered, it is a bit of a "chicken and egg" debate. Some cities decided not to replace or repair water fountains when they were broken or worn out because of the costs involved, and also because they assumed (correctly) that most people would be bringing their own bottled water.

We are happy to report that there has been a move in some countries—though sometimes a little humiliating—to revive the water fountain. In London, the first fountain to be built in Hyde Park in 30 years was unveiled in September 2009. The fountain cost a little over £30,000 (US $45,000) and was paid for by a private benefactor who was a trustee of the Royal Parks Foundation. Despite this somewhat over-the-top beginning, the nonprofit, charitable organization Drinking Fountain Association is working with city officials to establish 150 additional fountains in London's parks, at a small fraction of the cost of the Hyde Park fountain. Similar efforts are underway in other cities.

If Paris is the "city of lights," Rome is the "city of fountains." It has over 2,500 drinking fountains bringing pure, free water from Italy's Apennine Mountains to locations throughout the city. These drinking fountains are so much a part of the fabric of life in the city that, recently, there has been an effort to provide maps showing all the fountains' locations so that locals and tourists alike can get a drink.

And while we don't know what happened to our old war-surplus canteens, there is also a move to bring back the metal container, in the wake of recent health concerns over the safety of plastic, namely, that chemicals in some types of plastic may begin to leach potentially toxic chemicals into the water after weeks or months of use. Specifically, the plastic bottles may contain Bisphenol A (BPA), an estrogen-like chemical. In March 2010, the Environmental Protection Agency announced that it shares the concern of the Food and Drug Administration about the potential health impacts of BPA. Fortunately, metal containers, available in both aluminum and stainless steel, have made a comeback and can be readily found in stores.

REINING IN THE BOTTLER

The stories we have just told chronicle efforts to encourage the use of tap water. The following stories are about people in small rural towns who took steps to prevent bottled water companies from trying to take control of the local waterways.

The little Northern California town of McCloud was a prime candidate to become the home of a water bottling plant. The town offered access to a wonderful water source and they were in need of jobs.

In 2003, Nestlé Waters North America, the largest bottling company in the United States, discovered the clear, clean water in the upper reaches of the McCloud River, on the edges of Mount Shasta in Northern California. The company was fortunate: It had entered into an agreement with town officials that allowed it to draw water from the river for up to a hundred years, and there was no cap on the amount of water that it could pump from the river.

The contract was signed by the town manager (it was later described as a "behind closed doors" deal), and within months McCloud residents and local environmental groups began to express concern over the potential size of the project. They formed the "Protect Our Waters" coalition, which included the McCloud Watershed Council, California Trout, and Trout Unlimited. McCloud, though remote, is a popular destination spot for trout fishing. The coalition raised questions and highlighted flaws in the environmental review process that had allowed the bottling plant facility to come in. The dispute drew the attention of California's attorney general, Jerry Brown, who submitted a letter to the local planning authorities citing the deficiencies in the Draft Environmental Impact Report (DEIR) that was used to support the project's approval. Specifically, the attorney general's letter found that the DEIR, among other things, failed to address in any meaningful way the project's likely environmental impacts.

By 2008, with organized objections from local residents and statewide organizations, Nestlé agreed to scale back its plans for the plant and also to conduct additional scientific studies on its environmental impact on the river. In 2009, Nestlé decided to withdraw completely from McCloud and move its bottling operation 80 miles to the south to Sacramento, in a significantly scaled-down version. The plant will utilize Sacramento's tap water, but only 80 million gallons a year instead of the planned McCloud draw of 521 million gallons a year.

McCloud's strong resistance to Nestle was successful, but it was tame compared to the reaction of Bundanoon, New South Wales, Australia, which voted to ban the sale of bottled water completely. The town's merchants replaced the bottled water on their shelves with reusable bottles to be sold for the same price as bottled water. Customers can fill the reusable bottles for free in the public fountains.

Bundanoon's stance against bottled water grew out of its ongoing battle with the Australian water-extraction company, Norlex. Norlex had a license to extract groundwater from Bundanoon that dated back to 1995; it then sold the water to bottling companies. By 2006, the company was ready to move forward with its extraction plans, and the town went to court to rescind, or at least restrict, Norlex's license. The legal battle with Norlex continues, but Bundanoon made quite a statement by banning bottled water. In the words of the town's organizers: "We're hoping it will act as a catalyst to people's memories to remember the days when we did not have bottled water."

TAP PROJECT: PUTTING IT
ALL INTO PERSPECTIVE

In 2007, UNICEF started a program called the Tap Project in order to raise money and awareness about the plight of children who lack access to safe drinking water. The project invites diners at participating restaurants to donate one dollar or more for a glass of tap water,

> *A dollar or more is donated for a glass of tap water at participating restaurants. For each donation, UNICEF can provide one child with access to safe, clean water for 40 days.*

which would normally be free. Since then, the program has grown from several dozen restaurants in New York City to over 1,500 restaurants throughout the United States. With 3,000 volunteers in 40 cities, it is the largest volunteer effort UNICEF has ever done in the United States. It shows no signs of slowing down as plans to expand the program continue.

"The lack of sanitary and accessible water is one of the leading causes of preventable death in children throughout the world, and access to clean tap water is something Americans often take for granted," said Caryl Stern, president and CEO of the U.S. Fund for UNICEF.

Stern is right, and we don't mean to suggest that drinking bottled water is sinful and that we must cease and desist. The popularity of water as a beverage represents a positive health trend. We think it is great that sales of bottled water are catching up to those of bottled sugar drinks. But there is no question that we need to refocus our priorities. While we are spending billions on bottled water, we have blinders on when it comes to the systems essential to our everyday well-being. As we seek out designer water, our overwhelmed water supply and sewer systems continue to pollute the rivers and, in some cases, poison us.

CONCLUSION

SO, NOW WHAT?

WE MUST TAKE ACTION.

We know we are on a collision course between the demands of increasing population and dwindling water sources. Climate change is a dangerous catalyst in this scenario. Can we avoid this collision and its devastating affects? Yes, but we must take action: First, we must use water more efficiently and productively. Orange County's water reuse was a lesson in water efficiency. Farmers in the Murray-Darling River Basin learned to make their scarce water more productive. Second, we must factor in the environmental costs. An example is a wastewater treatment plant that cleans the wastewater before discharging it into a river or lake, and produces more energy than it uses. Third, because this is an emerging crisis, we need to be prepared to continually strive for higher standards. The trick is to progress from where we are now to a world of sustainable allocation. In this book, we have described some successful transitions that have moved us closer to that goal.

WE NEED COLLECTIVE ACTION.

The water consumer must be a partner in these problem-solving efforts because making the transition requires collective action. We are more likely to accomplish our goals when we do things together, as opposed to having things done to us. The success story in Brasilia was only possible with collective action. San Francisco could only begin to repair its dilapidated sewers with support from its voters.

The solutions profiled in this book can be applied right now to our water crisis. Although we discuss cases from around the globe, there might be a charge of chauvinism when we discuss global issues from a U.S. perspective. Sometimes that charge may be appropriate, but for water issues it is less so. The reason is the United States is a mini-continent that contains every possible type of water regime existing around the globe—from rain forests to well-watered plains, from glaciers and ice-fields to some of the world's hottest and driest deserts. Moreover, in many aspects of practical water policy the United States, because of its historical economic and social development, has had to face these challenges first. Now, as other countries of the world become increasingly affluent and developed—India and China are prime examples—all of the issues we face in the United States are now more evident around the globe. Though these situations may not exactly match the problems facing you or your community, we identified several principles that can be nearly universally applied. Let's recap some of the important lessons we've learned.

WATER IS A FINITE RESOURCE.
WE MUST MAKE IT LAST.

The technology is available to make significant improvements in stretching water further both in household and industrial or agricultural uses.

In the urban setting, Singapore's NEWater program recycled its sewer water to produce drinking water and water for high-tech industries. With available recycling technology, there are a myriad of ways to greatly increase water efficiency through water reuse.

The technology is available to increase efficiency in agricultual water use. Eugene Glock used readily available, relatively low-cost technology and put it to work to improve productivity and the total value of his crops. As a leader in Nebraska's farming community, he was able spread his innovative thinking throughout that community. Because of his leadership, the section of the Ogallala aquifer that provides water to Glock and his neighbors is less stressed than in other regions drawing from the same aquifer.

IMPROVE WATER SECURITY.
THE GREATEST RISK TO OUR
WATER SECURITY IS NOT TAKING ONE.

As we saw in Orange County, government leaders that were willing to take risks were able to achieve greater water security than their more cautious neighbors. Orange County is now in a much better position to guarantee water to its growing population than the surrounding counties that are still reliant on expensive and scarce imports from elsewhere.

East Bay Municipal Utility District in Oakland took a chance on a new method for treating wastewater and turned waste into a resource, not another source of water and air pollution. East Bay MUD runs its system more efficiently by producing power instead of just using it. And by using food scraps for its energy production, it also reduces methane gas, one of the most significant greenhouses gases.

On a larger scale, for countries involved in transboundary water conflicts, the greatest risk is not looking beyond your country's self-interest, whereas clever basin-wide collaboration could improve the productivity of water use.

ETERNAL VIGILANCE:
KEEP STRIVING FOR HIGHER STANDARDS.

St. Petersburg, Florida, broke new ground by becoming the first city in the United States to use recycled water for landscape irrigation. And, while the city's efforts have been very successful in significantly reusing water and replacing potable water with recycled water, they need to keep fostering that pioneering spirit to rein in the overuse of water. For example, they can employ incentives, like those used in a number of different cities across the country, to make their gardens more compatible with Florida's weather. To stay ahead of increasing populations, affluence, and climate change, we must plan for the future.

As part of reaching for higher standards, we must think holistically; connect the dots. The actions of some concerned city officials and personnel at the water and wastewater utilities in San Francisco and Santa Rosa are stellar examples of the kind of creative solutions we need to prevent the further pollution of our environment, the further depletion of our energy sources, and to make the best use of our water and wastewater.

We must maintain or retrofit our water and wastewater systems to make them as energy efficient as possible. At the bare minimum, these systems should be operated in ways that do not deplete our energy sources nor contribute to the problem. In San Francisco we saw the positive results of a public utility operating in a holistic manner. San Francisco reduced the problems of grease-clogged sewers that were causing dangerous sewer overflows and turned the grease into a clean, renewable source of biodiesel, which will benefit the water we drink as well as the air we breathe. Similarly, Santa Rosa's sewage now powers California's largest producer of renewable energy—more than solar and wind energy combined—at the Geysers geothermal plant in Geyserville. Santa Rosa's water utility did more than just stop polluting the Russian River; it turned its sewage into a renewable energy source.

TO CHANGE BEHAVIOR, PROVIDE INCENTIVES.

Regulatory Incentives

As we have seen, the farming community does not always make water efficiency its first priority. It needs to be reminded that water resources are limited, and, in some instances, as we saw in California's Imperial Valley, legislative or regulatory enforcement provides the needed incentive to operate more efficiently.

One of the lessons to be learned from the Imperial Valley case is that federal officials had to take the lead. Unfortunately, it can be difficult to say who is in charge of water issues in the United States because there are over a dozen federal departments and agencies that manage water, wastewater, and related issues, such as energy. While water scarcity may pose a greater threat to national security than energy shortages, water does not command the same level of federal attention as energy. Why isn't there a federal cabinet post or key government agency focused only on water?

Economic Incentives

Water is priceless, especially if you don't have it or are running out of it. Most of us think of water as a free public good because it falls out of the sky. But if we want to use this precious resource more efficiently, we need to give it a monetary value. In Boston, the economic incentive of steep rate increases, coupled with conservation efforts, led to a marked increase in water use efficiency.

In the agricultural sector, economic incentives do work, as we saw in the Murray-Darling River Basin. There is an old saying in the water business: "Water flows uphill to money." In Australia, water may not have traveled uphill, but the money moved the farmers to adapt more efficient methods of farming and to allocate their water to higher-value, less water-intensive crops.

Economic incentives and financial investment can also play a role on a broader scale in achieving cooperation among nations. Investments and economic incentives by donor nations were essential in reaching compromises in transboundary water conflicts.

POVERTY SHOULD NOT BE AN IMPEDIMENT TO PROVIDING BASIC SERVICES.

The infrastructure needed to protect our public health and our environment is expensive to build and to maintain. We must be willing to make this investment in our environmental future while, at the same time, always working to maximize cost effectiveness and affordability. Our cases from Brazil and Pakistan show that poverty need not be a barrier to providing basic services to save lives, reduce misery, and protect the environment. As we saw in those examples, basic infrastructure can be provided to poorer communities through innovative and thoughtful leadership. Water and sewer systems can be provided at a fraction of the cost with community involvement and "sweat equity." Literacy levels should be no impediment to informing and involving the consumer.

EVERYBODY INCLUDED:
THE CONSUMER MUST BE INVOLVED.

Real progress will require real involvement. It will require our encouragement, approval, money, and, for some of us, lifestyle changes. Do we really want to poison ourselves or the planet with our own waste? Do we really want to pay exorbitant prices for food or suffer food scarcities?

We need to rearrange our priorities. Those who live in the developed countries around the globe are very lucky to have easy, affordable access to clean, safe water. Let's take advantage of this and pour some of the billions we spend on bottled water back into our infrastructure. The energy drain and environmental harm that results from our collective infatuation with bottled water should not be a part of our water crisis or the environmental one.

Civic leaders and water utilities also have a responsibility to let community residents and businesses know that they have systems that are working for them 24/7. The San Francisco sewer case demonstrates the positive effects of including the consumer. Orange County's success in recycling water was a stark contrast to San Diego's political drama, and Orange County's success was dependent on consumer acceptance. As responsible citizens, we should be asking, What's up with our most essential infrastructure? And if your water utility or civic leaders aren't letting you know, *ask them:* Send them an e-mail along with a copy of your water and sewer bill and ask how the systems are working. Are they being adequately maintained, and is the community getting a status report?

We began this book describing a looming crisis. We examined a wide variety of actions and policies that, taken together, could solve this crisis. Consider the impact of a 10 percent improvement in agriculture's water use; such a minor improvement would make a significant change in water availability. Or consider the implementation of inexpensive technologies to provide safe drinking water and sanitation that result in improved productivity and public health. The widespread use of intervention to solve this crisis ultimately relies on the leadership of politicians and opinion leaders combined with grassroots demand for action. Your involvement is needed for something more than sustainable water use. It is more basic: Your involvement is needed to assure water for survival.

ACKNOWLEDGMENTS

We are forever grateful for the encouragement and support of Suzanne Ogden and Susan Hirsch. We are very appreciative of the assistance of our colleagues Margaret Owens and Shelly London and the opportunities provided by the Harvard Advanced Leadership Initiative and its leader Rosabeth Moss Kanter. Of course, this book was only possible with the help of our editor Luba Ostashevsky and her colleagues.

This book benefited from so many in so many ways, we cannot thank them all personally, but we do want to particularly thank the following:

Phil Anthony, Steve Ashley, Jared Blumenfeld, John Briscoe, Michael Carlin, Dan Carlson, Olivia Chen, Gina DePinto, Dennis Diemer, Stephen Estes-Smargiassi, Bob Fisher, Eugene Glock, David Grey, Laura Haight, Michael Hanemann, Ed Harrington, Laura Harnish, Lewis Harrison, Nagaraja Rao Harshadeep, Janet He, Annette Huber-Lee, Mary Hughes, Tyrone Jue, Jamie Kaplan, Aki Kawaski, Kevin Kelly, Harlan Kelly, Jeff Kightlinger, Angela Licata, Julie Malakie, Ed McCormick, Curtis McKnight, Robert Meany, Charles Miranda, Tommy Moala, David Molden, Mike Muller, Pat Mulroy, Roger Patterson, John Riera, Mike Sangiacomo, Hayden Simmons, Maureen Stapleton, Michelle Stolte, Eleanor Torres, Samir Toubassy, Tony Winnicker, Kary Ving, Kheng Guan Yap and Mike Young.

BIBLIOGRAPHY

The bibliography is arranged by chapter; entries are listed in the order in which they were used.

CHAPTER ONE

Faures, J.-M. "Indicators for Sustainable Water Resources Development." Food and Agriculture Organization of the United Nations, 1997. http://www.fao.org/docrep/w4745e/w4745e0d.htm (accessed May 10, 2010).

Hoekstra, Arjen Y. and Ashok K. Chapagain. *Globalization of Water: Sharing the Planet's Freshwater Resources.* Oxford: Blackwell Publishing, 2008.

Magee, Mike. *Healthy Waters: What Every Professional Should Know About Water.* Bronxville, NY: Spenser Books, 2005.

Doud, Gregg and Julie McWright. "U.S. Cattlemen Look to China for Market Growth." *National Cattlemen's Beef Association Issues Update,* May-June 2006. http://www.beefusa.org/uDocs/uscattlemenlooktochinaformarketgrowth.pdf (accessed May 10, 2010).

Rogers, Peter. "Facing the Freshwater Crisis." *Scientific American* (August 2008): 46–53.

UNESCO World Water Assessment Programme. *Water a Shared Responsibility.* The 2nd United Nations Water Development Report, 2006. http://www.unesco.org/water/wwap/wwdr/wwdr2/ (accessed May 14, 2010).

Fox, K. R. and D. A. Lytle. "Milwaukee's Crypto Outbreak: Investigations and Recommendations." *Journal of the American Water Works Association (AWWA)* 88 (1996): 87–94.

Department of Water Resources, State of California. "Climate Change in California." http://www.water.ca.gov/climatechange/docs/062807factsheet.pdf (accessed May 10, 2010).

Water Utilities Climate Alliance. "Options for Improving Climate Modelling to Assist Water Utility Planning for Climate Change." *White Paper,* December 2009. http://www.wucaonline.org/assets/pdf/actions_whitepaper_120909.pdf (accessed May 10, 2010).

The California Energy Commission. "California's Water-Energy Relationship." Final Staff Report CEC-700–2005–011-SF, November 2005.

NRDC Natural Resources Defense Council: The Earth's Best Defense. "Energy Down the Drain: The Hidden Costs of California's Water Supply, Executive Summary." http://www.nrdc.org/water/conservation/edrain/execsum.asp (accessed May 10, 2010).

U.S. Government Accountability Office (GAO). *Clean Water Infrastructure: A Variety of Issues Need to Be Considered When Designing a Clean Water Trust Fund.* Report to Congressional Requesters GAO-09–657, May 2009. http://www.gao.gov/new .items/d09657.pdf (accessed May 10, 2010).

CHAPTER TWO

California Department of Water Resources. *California Water Plan Update 2005.* http://www.waterplan.water.ca.gov/previous/cwpu2005/index.cfm (accessed May 12, 2010).

Dickinson, Ann. "Where Water Runs Uphill." *Quest—A KQED Multimedia Series Exploring Northern California Science, Environment and Nature*, June 5, 2008. http:// www.kqed.org/quest/blog/tag/harvey-o-banks-pumping-plant/ (accessed May 10, 2010).

U.S. Department of the Interior. "Delta Subsidence in California: The Sinking Heart of the State." United States Geological Survey FS-005–00, April 2000. http://ca .water.usgs.gov/archive/reports/fs00500/fs00500.pdf (accessed May 18, 2010).

Fimrite, Peter. "U.S. Issues Rules to Protect Delta Smelt." *San Francisco Chronicle*, December 16, 2008.

Sullivan, Colin. "California Water Agency Changes Course on Delta Smelt." *The New York Times*, May 12, 2009.

Archibald, Randal C. "From Sewage, Added Water for Drinking." *The New York Times*, November 27, 2007.

Stein, Joel. "O.C.'s Water is No. 1: Getting a Taste of Orange County's Toilet-to-Tap Program." *Los Angeles Times*, December 14, 2007.

Union-Tribune Editorial. "Yuck! San Diego Should Flush "Toilet-to-Tap" Plan." *The San Diego Union-Tribune*, July 24, 2006.

Orange County's Groundwater Authority. "News Release: Orange County Water District Beats Out over 400 International Nominees to Win the 2008 Global Water Award." Fountain Valley, CA: Orange County Water District, May 6, 2008.

Morrison, Patt. "L.A.'s Water Makeover." *Los Angeles Times*, July 24, 2008. http://articles.latimes.com/2008/jul/24/opinion/oe-morrison24 (accessed May 12, 2010).

PUB, Singapore's National Water Agency. "History of NEWater, 2010." http://www .pub.gov.sg/about/historyfuture/Pages/NEWater.aspx (accessed May 12, 2010).

Tigno, Cesar. "NEWater: From Sewage to Safe." Asian Development Bank, December 2008. http://www.adb.org/Water/Actions/sin/NEWater-Sewage-Safe.asp (accessed May 12, 2010).

Blizin, Jerry. "City's Water Future Ran Dry." *St. Petersburg Times*, November 17, 2009.

The City of St. Petersburg, FL. "Reclaimed Water." http://www.stpete.org/water/reclaimed_water/ (accessed May 10, 2010).

Vander Veldt, Jessica. "Something Greener Than Grass." *St. Petersburg Times*, August 5, 2009.

Segrest, Melissa. "Cash for Grass: Las Vegas Residents Get Rebates for Tossing Their Turf." *Green Right Now*, July 20, 2009. http://www.greenrightnow.com/boston/2009/07/20/cash-for-grass-las-vegas-residents-get-rebates-for-tossing-their-turf/ (accessed May 10, 2010).

Southern Nevada Water Authority (SNWA). "Water Smart Landscapes Rebate." http://www.snwa.com/html/cons_wsl.html?print=yes (accessed May 10, 2010).

Green, Emily. "The Dry Garden: L.A. Offers Rebate for Ripping Out Your Lawn." *Los Angeles Times*, June 10, 2009.

CHAPTER THREE

California Department of Food and Agriculture (CDFA). "California Agricultural Highlights 2008–2009." http://www.cdfa.ca.gov/statistics/files/AgHighlights-Brochure09.pdf (accessed May 11, 2010).

State Water Resources Control Board, State of California. "Decision Regarding Misuse of Water by Imperial Irrigation District, Decision 1600, June 1984." http://www.sci.sdsu.edu/salton/iidAllegedWasteofWater.html (accessed May 10, 2010).

State Water Resources Control Board, State of California. "Order 88–20 adopted on September 7, 1988."

The Imperial Valley Economic Development Corporation (IVEDC). "Quick Facts on Imperial Valley Agriculture." http://www.ivedc.com/?pid=737 (accessed May 11, 2010).

Sterngold, James. "San Diego Buying Water—and Its Freedom." *The New York Times*, August 6, 1996.

Perry, Tony. "Key Deal Near on Shift of Water to Cities From Farms." *Los Angeles Times*, December 12, 1997. http://articles.latimes.com/1997/dec/12/news/mn–63301 (accessed May 10, 2010).

Perry, Tony. "Manager of Imperial Water District is Fired." *Los Angeles Times*, January 7, 1999.

Photographs from the Farm Security Association (FSA). "Migrant Workers" by Dorothea Lange (photographer). http://memory.loc.gov/ammem/fsahtml/fachap03.html (accessed May 10, 2010).

Norton, Gale A. "Speech to the Colorado River Water Users Association by the Honorable Gale Norton, Secretary of the Interior, Las Vegas, Nevada, December 16, 2002." http://www.interior.gov/news/norton1.html (accessed May 11, 2010).

Murphy, Dean E. "Southern California Water Officials Race Deadline of Tonight." *The New York Times*, December 31, 2002.

Gardner, Michael. "Imperial Decision." *SignOnSanDiego.com by The San Diego Union-Tribune*, December 8, 2002. http://web.signonsandiego.com/news/reports/water/20021208–9999-water.html (accessed May 10, 2010).

"Interior Secretary Cuts California's Share of Colorado River Water." *U.S. Water News Online*, December 2002. http://www.uswaternews.com/archives/arcsupply/2int sec12.html (accessed May 11, 2010).

Yniguez, Rudy. "IID Board Votes 3–2." *Imperial Valley Press Online*, October 3, 2003. http://www.ivpressonline.com/articles/2003/10/03/news/top_story/news01.txt (accessed May 11, 2010).

Glenn, Megan. "Imperial Irrigation District to Meet Water Transfer Requirements by Stopping Canal Spills." *Imperial Valley Press Online*, July 28, 2009. http://www.iv pressonline.com/articles/2009/07/29/local_news/news07.prt (accessed May 10, 2010).

Gleick, Peter H. *The World's Water 2008–2009*. Washington, DC: Island Press, 2009.

Murray-Darling Basin Authority, Australian Government. "Overview of Permanent Interstate Water Trading." Fact Sheet 1, May 2006. http://publications.mdbc.gov .au/download/PIWT%20Fact%20Sheet%201.pdf (accessed May 11, 2010).

"Australia's Water Shortage: The Big Dry." *The Economist*, April 26, 2007.

Murray-Darling Basin Authority, Australian Government. "A Report from the CSIRO to the Murray-Darling Basin Authority: Advice on Defining Climate Scenarios for Use in the Murray-Darling Basin Authority Basin Plan Modeling." MDBA Technical Report Series: Basin Plan: BP01, July 2009. http://www.mdba.gov.au/files/ publications/Defining-climate-scenarios-report-from-CSIRO.pdf (accessed May 11, 2010).

"South Australia's Water Shortage: In Need of a Miracle." *The Economist*, May 7, 2009.

Molden, David, ed. *Water for Food, Water for Life: A Comprehensive Assessment of Water Management in Agriculture*. London: Earthscan, 2007.

CHAPTER FOUR

Redman, Ryan L., Cheryl A. Nenn, Daniel Eastwood, and Marc H. Gorelick. "Pediatric Emergency Department Visits for Diarrheal Illness Increased After Release of Undertreated Sewage." *American Academy of Pediatrics* 120 (December 2007):e1472-e1475. http://www.pediatrics.org/cgi/content/full/120/6/e1472 (accessed May 11, 2010).

Dorfman, Mark. "Swimming in Sewage: The Growing Problem of Sewage Pollution and How the Bush Administration is Putting Our Health and Environment at Risk." National Resources Defense Council (NRDC), February 2004.

"Milwaukee's Sewage Problems Highlight National Issue." *U.S. Water News Online*, July 2004. http://www.uswaternews.com/archives/arcquality/4milwsewa7.html (accessed May 11, 2010).

Duhigg, Charles. "Saving U.S. Water and Sewer Systems Would Be Costly." *The New York Times*, March 14, 2010.

U.S. Environmental Protection Agency (EPA). "Preventing Waterborne Disease: A Focus on EPA's Research." http://www.epa.gov/ord/NRMRL/pubs/640k93001/ 640k93001.pdf (accessed May 11, 2010).

Lautenberg, Frank R., Senator. "Press Release of Senator Lautenberg on September 17, 2008: Senate Panel Approves Two Lautenberg Bills to Improve Water Quality and Safety . . ." http://lautenberg.senate.gov/newsroom/record.cfm?id=303121 (accessed May 11, 2010).

Duhigg, Charles. "As Sewers Fill, Waste Poisons Waterways." *The New York Times*, November 23, 2009.

UN International Year of Sanitation 2008. "Global Experiences in Improving Sanitation and Hygiene: What is Working and Where?" http://esa.un.org/iys/docs/6.%20Global%20Experiences.pdf (accessed May 11, 2010).

The World Bank. *World Development Report 1992: Development and the Environment.* Chapter 2, Environmental Priorities for Development, 44–63. New York: Oxford University Press, 1992.

Melo, Jose Carlos. "The Experience of Condominial Water and Sewerage Systems in Brazil: Case Studies from Brasilia, Salvador and Parauapebas." New York: The World Bank, August 2005. http://siteresources.worldbank.org/INTWSS/Resources/CondominialWater–84.pdf (accessed May 11, 2010).

Barreto, Mauricio L., Bernd Genser, Agostino Strina, Maria G. Teixeira, *et al.* "Effect of City-Wide Sanitation Programme on Reduction in Rate of Childhood Diarrhoea in Northeast Brazil: Assessment by Two Cohort Studies." *The Lancet* 370 (November 10, 2007): 1622–1628.

"Orangi Pilot Project. "Low Cost Sanitation Programme." Research and Training Institute OPP-RTI. http://www.oppinstitutions.org/athespprogram.htm (accessed May 11, 2010).

Ali, Mir Hussain. "Orangi Pilot Project." In *Proceedings of Kitakyushu Initiative Seminar on Public Participation.* http://kitakyushu.iges.or.jp/docs/mtgs/seminars/theme/pp/Presentations/2_Water/Karachi.pdf (accessed May 11, 2010).

Yap, Kheng Guan. "From Zero to Hero: NEWater Wins Public Confidence in Singapore." Presented at Singapore International Water Week, June 2009.

CHAPTER FIVE

Brabeck-Letmathe, Peter. "A Water Warning." *The Economist* 00130613 (December 20, 2008).

Smith, Adam. *Inquiry into the Nature and the Causes of the Wealth of Nations.* Originally published in 1776.

Hannemann, W. Michael. "The Economic Conception of Water." In *Water Crisis: Myth or Reality?*, edited by Peter Rogers, M. Ramon Llamas, and Luis Martinez-Contina, Chap. 4, 61–91. London: Taylor & Francis plc, 2006.

Rogers, Peter, Radhika deSilva, and Ramesh Bhatia. "Water is an Economic Good: How to Use Prices to Promote Equity, Efficiency, and Sustainability." *Water Policy* 4 (2002): 1–18.

U.S. Environmental Protection Agency (EPA). "Measuring the Benefits of Water Quality Improvements Using Recreation Demand Models: Part I, Policy Planning and Evaluation." EPA-230–10–89–069, October 1989. http://yosemite.epa

.gov/ee/epa/eerm.nsf/vwAN/EE-0004B-1.pdf/$file/EE-0004B-1.pdf (accessed May 14, 2010).

Finnegan, William. "Leasing the Rain: The World is Running Out of Fresh Water, and the Fight to Control It Has Begun (Letter from Bolivia)." *The New Yorker* 78 (April 8, 2002): 43–53.

Martin, William E., Helen M. Ingram, Nancy K. Laney, and Adrian H. Griffin. *Saving Water in a Desert City.* Washington, DC: Resources for the Future, 1984.

Dolin, Eric Jay. *Political Waters: The Long, Dirty, Contentious, Incredibly Expensive, but Eventually Triumphant History of Boston Harbor—A Unique Environmental Success Story.* Amherst, MA: University of Massachusetts Press, 2001.

Haar, Charles M. *Mastering Boston Harbor: Courts, Dolphins, and Imperiled Workers.* Cambridge, MA: Harvard University Press, 2005.

Massachusetts Water Resources Authority (MWRA). *Annual Water and Sewer Retail Rate Survey: December, 2008.* The Community Advisory Board to the MWRA, December, 2008. http://www.mwra.state.ma.us/ (accessed May 14, 2010).

CHAPTER SIX

U.S. Environmental Protection Agency (EPA). "Controlling Fats, Oils, and Grease Discharges from Food Service Establishments." National Pretreatment Program, Office of Water, EPA-833-F-07–007, July 2007. http://www.epa.gov/npdes/pubs/pretreatment_foodservice_fs.pdf (accessed May 11, 2010).

Hennessy-Fiske, Molly and Tami Abdollah. "Summer Bummer in Long Beach: Despite Efforts to Keep Waters Clean, a Sewage Spill Forces the Year's Fourth Shore Closure." *Los Angeles Times*, July 29, 2008.

U.S. Department of Justice. "Tyson Pleads Guilty to 20 Felonies and Agrees to Pay $7.5 Million for Clean Water Act Violations." http://www.justice.gov/opa/pr/2003/June/03_enrd_383.htm (accessed May 11, 2010).

Humphries, Jodie. "The Impact of Domestic Food Waste on Climate Change." *Next Generation Food*, March 4, 2010. http://www.nextgenerationfood.com/news/looking-at-food-waste/ (accessed May 11, 2010).

Duhigg, Charles. "Health Ills Abound as Farm Runoff Fouls Wells." *The New York Times*, September 18, 2009.

"Foster Farms Receives California's Highest Environmental Honor—the Governor's Environmental and Economic Leadership Award; Company Is Also the Recipient of the 'POWER Award,' Both for Voluntary Solution to Complex Disposal Issues." *Business Wire*, November 23, 2005.

U.S. Environmental Protection Agency (EPA). "Turning Food Waste into Energy at the East Bay Municipal Utility District (EBMUD)." http://www.epa.gov/region9/waste/features/foodtoenergy/ (accessed May 13, 2010).

Hall, Kevin D., Juen Guo, Michael Dore, and Carson C. Chow. "The Progressive Increase of Food Waste in America and Its Environmental Impact." *PLoS ONE* 4 (November 2009):e7940 1–6. http://www.plosone.org/article/info%3Adoi%2F10.1371%2Fjournal.pone.0007940 (accessed May 13, 2010).

California Energy Commission. "Santa Rosa Geysers Recharge Project: GEO-98–001." Final Report, October 2002. http://www.energy.ca.gov/reports/2003–03–01_500–02–078V1.pdf (accessed May 11, 2010).

McCoy, Mike. "The Geysers Getting Credit for a Cleaner Russian River." *The Press Democrat*, May 26, 2009. http://www.pressdemocrat.com/article/20090526/news/905261005 (accessed May 13, 2010).

CHAPTER SEVEN

Gleick, Peter, quoted from http://www.sfgate.com/cgi-bin/blogs/gleick/detail?entry_id=40585 (accessed May 17, 2010).

Wolf, A. T., J. Natharius, J. Danielson, B. Ward, and J. Pender. "International River Basins of the World." *International Journal of Water Resources Development* 15 (1999): 387–427.

Salman, S. M. A. "International Watercourses: Enhancing Cooperation and Managing Conflict." World Bank Technical Paper No. 414, 1998.

Le Marquand, David G. *International Rivers: The Politics of Cooperation*. Vancouver: Westwater Research Center, University of British Columbia, 1977.

Rogers, Peter. "The Value of Cooperation in Resolving International River Disputes." *Natural Resources Forum* (May 1993): 117–131.

Gulhati, Niranjan Das. *Indus Waters Treaty: An Exercise in International Mediation*. Foreword by Sir William Iliff. Bombay, New York: Allied Publishers, 1973.

Nile Basin Initiative. http://www.nilebasin.org/index.php?option=com_content&task=view&id=13&Itemid=42 (accessed May 14, 2010).

Mekong River Basin Commission. "State of the River 2010." Vientiane, Laos, April 2010.

Osborne, Milton. *The Mekong River Under Threat*. Sydney, Australia: Lowy Institute for International Policy, 2009.

Fisher, Franklin M., Annette Huber-Lee, et al. *Liquid Assets: An Economic Approach for Water Management and Conflict Resolution in the Middle East and Beyond*. Washington, DC: Resources for the Future, RFF Press, 2005.

CHAPTER EIGHT

Beverage Marketing Corporation of New York. "Bottled Water Perseveres in a Difficult Year, New Data from Beverage Marketing Corporation Show." News Release, April 20, 2009.http://www.beveragemarketing.com/?section=pressreleases (accessed May 17, 2010).

Vedantam, Shankar. "What's Colorless and Tasteless And Smells Like . . . Money?" *The Washington Post*, June 30, 2008. http://www.washingtonpost.com/wp-dyn/content/article/2008/06/29/AR2008062901872.html (accessed May 17, 2010).

Snitow, Alan and Deborah Kaufman. *Thirst: Fighting the Corporate Theft of Our Water*. San Francisco: Jossey-Bass, A Wiley Imprint, 2007.

U.S. Government Accountability Office (GAO). "Bottled Water: FDA Safety and Consumer Protections Are Often Less Stringent Than Comparable EPA Protections for Tap Water." Report GAO-09–610 to Congressional Requesters, June 2009. http://www.gao.gov/products/GAO-09–610 (accessed May 17, 2010).

Ness, Carol. "Local Tap Water Bubbles Up in Restaurants." *San Francisco Chronicle*, March 21, 2007.

Credeur, Mary Jane and Thomas Mulier. "Nestle Loses Sales as Alice Waters Bans Bottled Water (Update3)." Bloomberg.com, January 22, 2008. http://www.bloomberg.com/apps/news?pid=20670001&sid=aeedkn1t3aus (accessed May 17, 2010).

Fortson, Danny. "Bottle vs. Tap Grudge Match Hots Up: Sick of Being Accused of Harming the Environment, the Bottled-Water Companies Are Fighting Back with Campaigns Attacking Tap Water." TimesOnline, June 7, 2009. http://business.timesonline.co.uk/tol/business/industry_sectors/retailing/article6446066.ece (accessed May 17, 2010).

Low, Valentine. "£30,000 Hyde Park Fountain Aims to Sink Bottled Water Craze the Hyde Park Fountain." Published on Polaris Institute. http://www.polarisinstitute.org/print/599 (accessed May 17, 2010).

Gallagher, Paul. "London's New Drinking Fountains a Challenge to Bottled Water Industry." *The Observer*, October 4, 2009. http://www.guardian.co.uk/environment/2009/oct/04/london-drinking-fountains-water-industry (accessed May 17, 2010).

U.S. Environmental Protection Agency (EPA). "EPA to Scrutinize Environmental Impact of Bisphenol A." EPA News Releases by Date. Release date March 29, 2010. http://yosemite.epa.gov/opa/admpress.nsf/2010%20Press%20Releases!OpenView&Start=199 (accessed May 17, 2010).

California Trout, Inc. "Op-Ed: California Trout: Nestle's Leaving McCloud." YubaNet.com, September 14, 2009. http://yubanet.com/regional/Op-Ed-California-Trout-Nestle-s-Leaving-McCloud_printer.php (accessed May 17, 2010).

Brown, Edmund G., Jr. "Nestle Waters North America Environmental Impact Report." Letter to Terry Barber, Interim Planning Director, Siskiyou County Planning Department, Yreka, CA, July 28, 2008.

"NSW Town of Bundanoon Votes to Ban Bottled Water." News.com.au, July 9, 2009. http://www.news.com.au/nsw-town-of-bundanoon-votes-to-ban-bottled-water/story–0–1225747578818 (accessed May 17, 2010).

NY Public Interest Research Group (NYPIRG), et al. "Groups Cheer Passage of Bigger Better Bottle Bill: Update of State's Bottle Recycling Law Hailed as Major Environmental Victory." News Release, April 3, 2009. http://www.nypirg.org/enviro/bottlebill/BBBBpasses.pdf (accessed May 17, 2010).

Elliott, Stuart, "A Campaign for Clean Drinking Water Expands," *New York Times*, March 4, 2009, http://www.nytimes.com/2009/03/05/business/media/05adco.html

INDEX